REPORT

OF THE

NORTH-CAROLINA GEOLOGICAL SURVEY

AGRICULTURE OF THE EASTERN COUNTIES;

TOGETHER WITH

DESCRIPTIONS OF THE FOSSILS OF THE MARL BEDS.

Illustrated by Engravings.

BY

EBENEZER EMMONS.

RALEIGH:

HENRY D. TURNER.

1858.

HOLDEN AND WILSON,
Printers to the State.

To His Excellency, Thomas Bragg,

Governor of North-Carolina:

Sir:

I am gratified that another opportunity is furnished me to express my obligations to your Excellency for the interest you still entertain for the Geological Survey of North-Carolina. This fact, while it has been extremely gratifying, serves at the same time to impress me with the importance of the work, and to excite a fear, also, that it may fall short of your expectations, and thus disappoint, not only yourself, but many others who feel and manifest an interest in its success. No one, however, could feel a greater disappointment at such a result than myself; and fearing that my labors, together with the labors of those who assist me, might fail to be satisfactory, I have certainly lost no time, nor spared any work, which I deemed necessary to secure the wished-for result.

With the consciousness, then, of having done this much for its success, I submit with cheerfulness this second report to your Excellency's consideration.

I am, Sir,

Your obedient servant,

EBENEZER EMMONS.

Raleigh, *March* 1, 1858.

PREFACE.

The subjects which are treated of in this Report, are mostly practical, and it has been my aim so to treat them, that the matter shall be useful. The agricultural part embraces descriptions and statements of the composition of many of the soils of the Eastern counties. These samples of soils which have been analyzed, are preserved in the Geological collection for future reference. I have sought to obtain all the practical information respecting them which I could, and for this end, the analyses have been usually carried as far as was necessary. The number of soils which have been thus submitted to analysis, are sufficient, probably, for the purposes intended by the projectors of the survey. I think they embrace all the classes of soils which exist in this section of the State. But there are, no doubt, many additional analyses, which would be useful where they appear to be special in their composition, and exhibit certain peculiarities. A class of soils of great interest exists in several of the eastern counties, of which a type is well known in the county of Hyde. I felt that it was an object to determine the composition of this class with accuracy, and to see it in place with the burthen of its crops still standing. In my researches, I have discovered that this peculiar soil exists in a greater or less degree of perfection in several other counties. In some instances, the soil is the same, but is less deep; in others, it is fully equal to the Hyde county or the Mattamuskeet lands, both in depth and richness. It seemed to be a prevailing impression that Hyde county soils existed no where else, and were confined to that county. But Onslow, Jones, Hanover, Brunswick, Beaufort, and others, still possess equally rich swamp lands.

The Gallberry lands, which occupy a middle position between these rich swamp lands and the sandy rolling uplands, are usually very poor; but there are many tracts which rank under this class, which may be cultivated profitably. There are two kinds of Gallberry lands: one which is black or blackish, which consists mainly of vegetable matter, and a white marine sand. This variety of this class is generally too poor to pay the expense of reclaiming. It may produce a few tolerably fair crops of corn, but it is soon exhausted, for it consists only of sand and vegetable matter. It may graduate into a better kind, as the white sand is exchanged for a drab colored one, and which becomes fine. The other variety of this class, is clay-colored, and is very stiff, and mixed with coarse particles of flint. It is almost impervious to water. It is naturally cold, and is not productive, prior to draining and the employment of fertilizers. It has a body, and is better than the black soil with the usual admixture of white sand.

In the examination of soils, the physical properties require as much attention as the chemical; for, in order that a good chemical mixture of elements may be fertile, they should possess a certain degree of adhesiveness or closeness, which will retain water. Those which are porous and coarse, permit water to pass through almost immediately. The result which follows, is fatal to plants, or crops of value; chemical action under those circumstances is too feeble to furnish it with sufficient nutriment. The fertilizers of the eastern and south-eastern counties have received all the attention which could be bestowed upon them. The great defect which I find in their composition is, the great excess of sand. This element being in excess, gives them only a local value; that is, they are not rich enough to permit of transportation to neighboring counties.

In order to increase their value, I have been led to entertain the opinion, that they may be washed. In this operation the sand may be separated from the valuable parts. This opinion, however, requires a confirmation by experiment. The material which remains after the sand is separated, contains phosphate of lime, carbonate of lime and magnesia, potash and soda; those elements which make the marl the most valuable. If any cheap process for washing the marls could be employed, the material could be transported to most of the midland counties with profit.

The cultivation of the grasses to a much greater extent than has hitherto been done in this State, has seemed to me very desirable. I have given considerable attention to the subject, and for the purpose of aiding, as far as possible, a measure of this kind, I have selected several of the most valuable for description, that information respecting their value, may be more widely spread. I am confident that many of them will succeed. Very few efforts have hitherto been made for their cultivation,—most planters entertaining the belief that it is impracticable, or else the labors of the plantation are supposed to be much more profitably directed to the raising of cotton. Under the present system of curing the grasses for winter fodder, tho labor is so much cheapened that it seems to me that the raising of cotton or any other of the great staples, will not interfere with the project of keeping more stock, and in a better condition than has hitherto been attempted in the State.

In connexion with the marls of the eastern counties, I have given a brief sketch of the fossils of the different kinds of beds. Those who will take the trouble to examine the figures of the fossils which belong to the different beds, will not fail to perceive the striking differences which prevail. It is, for instance.

exceedingly rare to find a species common to two beds, although they lie in juxtaposition; or one may repose upon the other. Hence, the utility of the presence of fossils to distinguish beds belonging to the different epochs from each other. Another object which I had in view in occupying so much space upon this subject, was, to aid those who wish to become acquainted with this interesting subject. Geology is now commanding the attention of some of the best minds in this country and Europe. It is invested with great importance and interest, as it is through the discoveries in this department of science, that we obtain a knowledge of the ancient history of the globe. This pursuit is especially recommended to the attention of the young. It will be found extremely interesting and useful, and no one will regret afterwards that he devoted a portion of his leisure hours to its study.

TABLE OF CONTENTS.

CHAPTER VIII.

CHAPTER IX.

CHAPTER X.

CHAPTER XI.

CHAPTER XII.

CHAPTER XIII.

CHAPTER XIV.

CHAPTER XV.

CHAPTER XVI.

CHAPTER XVII.

CHAPTER XVIII.

CHAPTER XIX.

INTRODUCTION.

It is one of the distinguishing characteristics of the day to attempt to utilize science. The leading minds of the age seem to be as intensely engaged in diffusing knowledge and disseminating it as common stock, as they are in acquiring it for themselves. The consequences which have already flowed from their efforts, are, to have already made knowledge relating to many departments the common property of the masses. This knowledge is not probably exact in many individuals, perhaps in none, excepting those who make those subjects objects of special study; but then, they know the nature of the subjects treated of, as well as many of the conclusions which have been obtained. They know enough to make intelligent inquiries, and a subject matter for conversation; their minds are sufficiently informed to lead them upon the proper road of inquiry. More than this has been gained in many instances in common life. The way is already prepared for a general diffusion of knowledge. Of those subjects which are the most useful to society, none occupy a higher rank than those which are related to agriculture. Thus, the chemistry of agriculture is of the highest importance. The mechanics of agriculture are also important, and more attention has been paid to this branch than the former. Indeed, one of the first evidences that agriculture was really upon the road of improvement, was the appreciation of better implements of husbandry. Their improvement was first attempted. It was right that it should be so, for to make chemical principles available at all, it was necessary to change by mechanical means the condition of the soil. Improvements, then, in agriculture, began at the right end. The more abstruse principles of the business have become subjects of investigation since, and now there are but few farmers who are entirely ignorant of the chemistry and other collateral branches of the philosophy of agriculture.

But still, these important hand-maids to this indispensable calling have only just begun to exercise an influence over old modes and old practices. But two great obstacles to the introduction of rational methods in agriculture are being rapidly removed; that is, prejudice in favor of the old methods pursued by the fathers, and prejudice against innovation. Whatever is good in the old methods will be retained; and ultimately, what is erroneous and worthless will be rejected. Improvements, however, in agriculture, are necessarily slow in getting a foothold; much more so than in the mechanic arts; for there are stronger prejudices to be overcome, and in the former it seems there is a ready appreciation of value in every improvement which is made, while in the latter, a prejudice is to be first overcome by ample experience. But we may be assured that, sooner, or later, the benefits of a change will appear, as the improvements address themselves to men's pockets, which is one of the most influential of motives in common life.

The principles which control industrial pursuits are perfectly simple; and being simple, have been and still will be liable to be overlooked. Who among the merchants of a village, acquires most rapidly, ease and independence for himself? It is the one who, from a more extensive acquaintance with his occupation, a more attentive observation of the markets, and a more careful application of his judgment, untiring energy and prudent industry, buys the best article and sells it the cheapest.

Who, among the mechanics of the town, commands the business in his special line of production? It is that one who has been thoroughly instructed in the principles of his handicraft, applies his mind and judgment to his labor, and by that means improves the articles he makes, or the modes of its manufacture, and can thereby outstrip his competitors by manufacturing more, as well as better, and selling cheaper. It is a natural result—a simple law of trade and commerce.

But who among the agriculturists of the land are the most prosperous? It is he who is not content to follow the beaten track of his forefathers, or pursue the course which they have pursued, and because they pursued and beat it, but he who thoroughly imbues his mind with sound principles, who studies into the nature of his processes, and the reason why he does this in preference to that; who investigates the nature of his soils, and fits them most perfectly for his crops, and is moreover seasonable in his preparation. He will raise the most to the acre, and have more to sell, and can sell the cheapest, and make the most money. The greatest production, coupled with the best, controls the pockets of the purchasers, and insures to him, what is ever sought after, the earliest independence and the first honors in the line of a profession.

What lies at the foundations of commerce? What spreads her sails, or generates the steam of our floating castles which ply from port to port and from country to country? It is agricultural production. There is no other substratum upon which the business of the world can rest. Nothing else can impel the mighty engines of commerce, or set in motion the locomotive, with its heavy train of cars. It is not because the merchant buys and sells again. That is not production. *But it is because the farmer produces.* The other is but a transfer, and is only an incident in trade. The production is the ruling cause. It is that which supports, which moves. Put a stop to production, and the wheels cease to move, the paddle ceases to turn, the locomotive stands still, and the whistle is no longer heard. Production is the great element of life in commerce and manufactures. It is because agriculture exists, that commerce thrives, that the merchant can buy and sell. The earth is properly called the common mother of all. Her fruits nourish us, and supply the materials for the arts and manufactures, and the articles for trade and commerce. The earth is the mother of all, but that does not justify the agriculturist in waiting for her fruits with folded arms, and to neglect to store his mind with the elements and principles of agricultural knowledge, or hope, in inactivity, on a good Providence, or good fortune. If mother earth is rightly depended upon, it will be accompanied by works and the study of principles as connected with what he is to do for his soils. He cannot ask much of mother earth, who neglects to study elements and principles in this connection. I say elements and principles, for it is not enough to know the mechanical part. It is not enough to know how to plow, and reap, and mow; these are a part of an education, but it is not all of it.

Thus, we see, the commanding position of agriculture. Its position is commanding, independent of the mode in which a community of individuals conduct it. As it regards this section of the Union, its importance increases with the population of our country. The Agriculturalist is not restricted to the production of bread. While her granaries are overflowing with corn and wheat, she has still two other great staples of trade to arouse her energy: *cotton* and *tobacco*. These have been and are increasing in importance from the day the first seed germinated in her soil. These are money crops. In all these great staples, industry need not be paralyzed, nor the spirits be made to sink for want of a market. No one needs fear that a surplus will be left on his hands; that his toils will be unrewarded or his industry avail him nothing. Such is the condition of the world, that the great staples are sought for from necessity. Cotton must be had at any price to satisfy the imperative wants of the world. The loom cannot stand still. The necessities of thousands now demand it. The force of habit in the use of tobacco is so strong and so general, that its price can never be less than it is now. It is rather probable that it will be higher. Its production may be cheapened, its cost may be diminished, but its price in market will never be less. The advantage will ever be on the side of the producer. Farming, then, has an advantage over all professions. There may be too many lawyers, physicians and merchants, but never too many farmers. This is so, because the seaports of the world are their markets, and because there is a world of human families which are not producers, and hence have to be fed, their looms kept running, and their habits gratified.

It is not, therefore, the domestic market which is to be supplied. The products of the soil of North-Carolina are consumed far away; some, in the cities of the north, but a far greater amount by the population of the Old World. Important measures are being taken abroad to supply cotton for English manufactories from India and Africa, and no doubt with the hope that, ultimately, this nation may make herself independent of this country with respect to this indispensable article. A project of this nature must be regarded with some concern. It cannot succeed immediately, and it is doubtful whether cotton can be produced in those countries, so as to compete successfully in market with our own. In the first place, the husbandry of cotton is fully understood in the Southern States; and in the second place, the adaptation of climate and soil is perfect, and the means for supplying fertilizers to sustain its continued production are equally well established. Marl is the true fertilizer for cotton. This is fully established by experience and chemical analysis. All these facts put it in the power of the South to sustain vigorously, for an indefinite term of years, its production. From the Roanoke to Florida, this fertilizer in numberless forms is inexhaustible. Hitherto, it has been almost impossible to be satisfied that there has been a systematic and sustained effort to carry this production to the limit which the want of it abroad demands. The time, however, has come, when its production has become doubly important. The hopes of foreigners for success in supplying themselves with cotton from India and Africa, are based in a good degree upon its failure here, through some misfortune, such as political revulsion, exhaustion of the soil and other casualties which may occur, but which cannot now be foreseen. As it regards the exhaustion of the soil, there need be no fear, with the resources at command. It is true that large tracts have been exhausted, but agriculture is understood better now than formerly; and hence, the planter is abundantly able to forestall such an event and prevent its occurrence.

But in any event, the principles stated in the foregoing paragraphs, will govern the market. The best and *cheapest* article will be bought, and that will insure its sale in any quarter of the globe, in spite of the combination of Cotton Associations to produce it in India and Africa. If American planters can produce the best at a lower rate than it can be produced in India, then American cotton will find a market in Liverpool. It is a simple question of production; and foreign efforts to secure a market and exclude the American cotton, will result simply in arousing the cotton planter to make a successful effort to retain his foothold in the market which he now supplies. When the cotton planting States have once fully taken into consideration their immense advantages for production, it seems impossible that they should sleep over them. Cotton, Indian corn, wheat and tobacco, four great staples on their hands, for which the markets of the world are open. These minor productions of the homestead furnish business for all. The Alleghanies and their slopes are well adapted to grazing, and hence the raising of stock will become an item of immense importance to planters. Intercourse with the extremes, the east and the west, will soon be made easy. It will be cheap, if an enlightened policy controls the fare upon railroads. If an opposite policy should unfortunately prevail, the hopes of the planter and graizer will be partially disappointed.

The encouragement for pursuing agriculture may be found in the certain prospect of the mining resources of the State. In the various branches of this business, it will ultimately be found, that a large population will have to be fed. A population devoted to this interest are not producers of bread, meat or fruits. They are necessarily dependent for all these and more; and hence, a home market is furnished, which, as far as it goes, is as important as the foreign.

But I need not dwell on the importance of agriculture; its importance is felt. I was more anxious in this connection, to state my views of an improved agriculture; one which is understood, or one founded upon established principles,—one which leaves a beaten road and inquires into the why and wherefore. This is the only kind of agriculture which will elevate the masses, and give laborers a *status* or standing beside professional men, and enable them to exercise an influence as wide as theirs. Regarded in this light, it is not simply an extraordinary crop, which is to be produced, but it is a development of the mental faculties. These are compatible objects. Indeed, they go almost necessarily together, because they are the result of an exercise of the mind. The labor of thinking is involved,—a labor which is not at first performed without effort,—for that reason many prefer to let others think for them; and hence, they continue in that unenviable condition which is properly called a *statu quo*.

RALEIGH, March 1, 1858.

REPORT

OF THE

NORTH-CAROLINA GEOLOGICAL SURVEY.

AGRICULTURE OF THE EASTERN COUNTIES.

PRELIMINARY REMARKS.

For any thing we know to the contrary, there is such an ample provision in the economy of nature, that the production of food shall not depend upon skill, or a deep acquaintance with the laws of life.

Seeds are sown broadcast, the winds waft them from their parent stocks, or they fall unheeded to their roots; yet such is the relation of seed to earth, air and moisture, that they germinate and become new individual plants which, in due time, contain the appropriate nutriment for some existing organism. It may be it is food only for the insect tribes, the beast of the field, or it may serve the table of the Prince.

The simple growth and nutriment of plants is independent of science, high culture, or skill in the ordinary round of nature.

There is a provision to meet a certain amount of the wants of life, so far as food is concerned, which may be obtained without tillage. It is, however, limited. When the habitations of men become concentrated upon a comparatively small area, or a dense population fills the land, the natural magazine which furnishes the ordinary or regular supply of nutriment to the vegetable, especially the cereals, then becomes insufficient to supply the increased demands of numbers, and hence the natural resources fail, and there ever afterwards exists a demand for skill and science to meet these artificial wants.

The first efforts to supply the meat and bread of a dense population, in the earliest stages of society, are those which belong to the simplest kinds. They consist mainly in providing more room, light and air, or providing for the free penetration of roots through the soil, and the exclusion of weeds or unnutritive plants. But, inasmuch as nutritive matter is measured out and limited, and as there is no special provision to create a new supply, constant removal will, in the course of years, so far diminish the original stock, that the plant ceases to grow or perfect its fruit, or does so under circumstances less favorable for its perfection.

At this period it becomes necessary to inquire how fertility, when lost, may be restored; and this inquiry becomes more pressing in the direct ratio that the population has increased.

Experience does, or may step in and postpone the period of exhaustion, and partially supply, for a time, the nutritive elements. But generally these shifts to postpone the period of exhaustion fail, for they are merely the efforts of the empyric. Empyricism in no business is likely to lead to the discovery of sound principles; indeed, it cannot inform us of the fact of exhaustion at all; and hence, empyricism is not in the direct road to improvement. In one instance it may prove successful, but in the many it fails; as it cannot assign a cause or state a reason.

The perfection of cultivation, or the perfection of agriculture, demands a reason; and the period when a reason can be assigned may be regarded as the third stage of improvement. It is at this stage that agriculture requires a direct inquiry respecting cause and effect; or, in other words, into antecedents and consequents, in order that it may make progress when the rules of empyricism fail. Agriculture, in some of its scientific aspects, has obscurities, because it has enquiries to make which are closely related to those of life; and life, whether regarded as a mysterious principle, or a force dependent upon chemical relations, or chemical actions, is profoundly mysterious. Calling this force life, without attempting to tell what it is, we know that it controls all the results effected in and by the vegetable tissues. An organ, as a whole, possesses no force: the leaf has no force, neither have the stems, bark or kernel. The force alluded to resides in the cell; and hence it is sometimes called a cell force, and the sum or aggregate force of all the cells of an organ secures all the results in their proper season. The matured fruit is the result then of the combined force of all the cells which compose it, acting under the influence of outward forces, as air, light and heat.

The sum or aggregate of these changes, however, from germination to the consumation of the mature fruit, is concealed from view. We know only the simple fact, that of change from day to day. Of the effective agency residing in the cell we know nothing. But fortunately the questions which belong to scientific agriculture have only a slight relation to these; they are not questions relating to cell force, or to life in the abstract. These are one step farther back than it is necessary to carry them. We need make no interrogatories respecting cell force, or life, in order to till the soil in the best modes, or to grow large crops of wheat. But still these obscure questions bear a relation sufficiently close to darken or cloud those which must be answered, and we almost instinctively pass from those investigations which lie in the field of research to those which are a step farther back, and lie beyond the limits of legitimate enquiry.

§ 2. The field of investigation is really much nearer to us and more within the scope of legitimate inquiry. If we wish to know what is the appropriate food of the wheat plant, we have only to analyze it, or to determine the elements which compose the kernel. It is not *how* it is made, *how* the cell power operates, but simply what the constituents of the wheat or corn plant are.

In practice, then, the farmer is merely required to sow his wheat upon grounds which contain enough of the elements it wants. It is true, certain collateral questions of great importance have to be answered, such as those which relate to the physical condition of the soil, the measures which ought to be adopted to prevent the operation of injurious agents,—as frost, drought, depredations of insects, etc.

When experiments and observation have satisfied the farmer respecting the composition of wheat, corn, and of the soil in which they are to be planted, he has only to secure the proper mechanical condition of the soil, and put it into that state which is best adapted to their constitution. From the foregoing statement, it is evident that the range of scientific enquiry is limited to an experimental circle. The farmer is not required to go out of that area to determine the true theory of agriculture, to perfect the art or the practical part of the business.

§ 3. The following report is based on the preceding views relative to the scope or range of agricultural enquiry. The planter or farmer may speculate on vital or chemical forces, and form such theories upon those recondite forces as best comport with his knowledge of facts and principles; yet, as has been said already, practical enquiries do not extend to them; it only demands a range of knowledge which is bounded by experimental researches, and the deductions which legitimately follow therefrom.

It is therefore true, that enquiries into the nature of the cell force or vital force are not excluded from the philosophy of vegetation, but these ultimate interrogatories have no practical utility, so far certainly as the principles of culture are concerned. From these remarks, however, it should not be inferred that agriculture requires only an extremely limited range of knowledge—that its connections with other sciences are distant and doubtful. So far is this from being true, that it may be shown that it is intimately related to, and dependent upon, several of the important branches of knowledge. We have seen, for example, how important chemistry is to agriculture. To this it is wholly indebted for its wonderful progress in modern times. Climatology also is closely related to agriculture, inasmuch as a knowledge of the influence of light and heat, air and winds, height and depth, must influence the farmer in his selection of crops for tillage, and the modes by which they should be treated. Soils too, being derived from rocks of different periods and constitutions, influence their composition and capabilities more or less. Close observation relative to those influences frequently establish important generalizations; and hence, geology may be regarded as a department very intimately connected with agriculture, and whose principles are capable of advancing its interests.

It is scarcely necessary to refer to botany, as an allied branch of science. A practical knowledge of soils may be derived from it. Nature rarely errs in collocation. Plants, without selecting soils in truth, do really flourish best on certain tracts whose soil is found to be adapted to their special wants. Some are lime, others are potash plants; and hence, the farmer may be satisfied where certain plants abound, that certain important constituents of soils are present.

Animals, however, form a large part of his care and oversight. Often his chief wealth consists in cattle. The rearing of stock of favorite breeds, their improvement in general, and often in special points, demands a knowledge of physiology and anatomy. There is property in a knowledge of the foot of the horse, the joints of the bullock and the structure of the hoof. There is property in a knowledge of the skull and the physiognomy of the horse and the kine; and there is property in the knowledge of habits and best food for cattle and flocks, and in the organization of the stomach and its dependencies.

The farmer and planter, therefore, may say that they have not only property in lands and in cattle, but also in the phenomena of nature, as they may make those phenomena subservient to their interests; the sunbeam and shade add golden dust to their stores, when seed times and tillage are chosen under the guidance of philosophy.

§ 4. While agriculture in all its aspects presents a wide field for investigation, it still has very clearly such subdivisions of labor, that in practice, it may reach a high degree of perfection. We find, for example, that climate frequently restricts the most profitable productions to one or two staples.

Cotton cannot be grown with profit north of Virginia. The sugar cane and coffee return profits only on our most southern border. Tobacco, though not so strictly limited by parallels of latitude, still requires certain peculiarities of climate and soil, which greatly restricts its cultivation. Tea requires a peculiar climate. In some parts of the world it rarely or never rains; in others, rains are frequent; in others still, there are seasons of rain followed by others which are rainless. These peculiarities favor the growth and perfection of a class or a family of plants, while, at the same time, others are excluded. Hence, the cultivation being limited, perfection in the culture of a few, necessarily reaches a better and higher grade of perfection, than if the attention of the planter was divided among many. Profit depends, in a great degree, upon the adaptedness of climate to a particular crop. The difference arising from the cultivation of a variety of cotton, which is perfectly matured in this climate, and one that does not attain perfectly that perfection, except under the most favorable circumstances, is very great in the long run. The rearing of cattle is much more profitable where they are at home, than where they require much attention and care to make them thrifty.

The cereals have the widest range, while plants of little value to man are often very restricted in their ranges. We recognize in this important fact, a prospective provision designed expressly for the benefit of man.

If the foregoing remarks are true, the education which agriculture demands, in order to improve its condition, requires that of the highest grade. Agriculture, while it is not to lose its place as an art, must, in order to advance, demand of its cultivators more knowledge of the collaterals. Some call this mere book learning which is of no account in practice; and in support of this view, say that agriculture has got along very well without them. Indeed none of our fathers had the benefit of the collateral or direct lights; and yet they made money by their simple modes of culture. This is no doubt true. The planters of North-Carolina found a rich virgin soil. The crops of maize required but little attention. Cotton at a later day became a profitable staple, its importance increased with the return of every year. But what have been the results upon the soil from the midland counties of North-Carolina to Alabama? Let one pass along the rail-road from Raleigh to Columbia, and then through Georgia to Montgomery. The exhaustion of the soil by its culture is too palpable and plain to be overlooked. Exhaustion on the whole route is the prominent feature. It took place slowly but surely. What were rich lands under the simple culture of the fathers, have now become the poor and worn out lands of their sons. It is at this stage that education or knowledge is demanded. The fathers got along very well, and made money; but the sons, though they may inherit money already made, must be content with that, or move away, or else seek by superior knowledge to replenish the worn out inheritance. New modes of culture must be devised, and a much greater amount of knowledge and skill will be required than the fathers possessed.

CHAPTER I.

Reference to a former report. The perfection of seed depends on the character of the soil. Nutrient matters necessary to animal life traced to the soil. Essential elements of a good. The soil the reservoir of all these elements. Character and classification of the soils in the Eastern counties. Importance of determining the smallest per centage of earthy matter in a vegetable soil, which is compatible with a remunerating crop. Some elements are more essential to form a good soil than others. The organs of a plant are composed of different elements. The extreme of certain kinds of soil. Remarks on the adaption of soils, together with a statement of their composition. Soil of the open ground prairie in Carteret county. Pocosin and swamp lands. Soils of Hyde county.

§ 5. In a former report, that of 1852, I deemed it necessary to point out certain facts which have a direct bearing upon the principles of agriculture, and which indeed appear to constitute the foundation upon which it is based; and as the present report may fall into the hands of those who may not have seriously reflected upon those principles, I now propose to recapitulate them very briefly.

Soils must contain a sufficiency of certain inorganic elements, otherwise no seed can be perfected. The elements which support animal life may be traced to those which exist in the vegetable, especially the seed and fruit. Hence, the important products of life are derived from the soil, it being possible to trace them back through the vegetable, and the reverse, from the soil through the vegetable to the animal. Those products of life then, which can be traced to no other source than the soil, must be regarded as essential elements of the soil, and as designed to sustain and support life. The office of the vegetable tissue through which they pass to fit them for sustaining animal life, are to simply modify, or to form new combinations, and not new substances or elements.

Those which I regard as essential to animal life, and all of which exist in the soil, are, phosphorus, sulphur, potash, soda, lime, magnesia, iron, silica, nitrogen, oxygen, hydrogen and carbon. They do not seem, in any instance, to enter into the composition of living bodies in the elementary state, but as compounds; thus hydrogen combines with oxygen and forms water, or nitrogen and forms ammonia; oxygen combines with phosphorus, sulphur, etc., before they are fitted to enter into the composition of the animal tissue.

The soil then, being the great reservoir or source of these elements which are truly essential to life, and so far as nutriment is concerned are dependent upon them, we cannot overestimate the importance of preserving it in the best condition; and when the soil is so far deprived of these elements that the crops are imperfect, we see the importance of those fertilizers which contain them. It appears also, that substances which do not contain them, have never been denominated fertilizers at all. Hence, when matters are added to soils, it is expected that they contain more or less of phosphorus, sulphur, potash, soda, etc., in certain states of combinations which the plant is able to obtain.

§ 6. The soils of North-Carolina are remarkable. They belong very frequently to the extremes of certain well distinguished classes. On the one hand, these extremes consist of sand, a marine product, nearly pure, or with only a trace of other matters; on the other, they are composed of nearly pure vegetable matter, with only a trace of earth or soil proper. These are not simply rare exceptions to the common run of soils, but they form classes. So also the stiff clays which are also marine deposits, form another class. These, however, do not materially differ in composition from the same class in other sections of the State.

The two former, I believe, are sectional, and are confined to the lower counties.

Besides the foregoing, where rocks exist near or at the surface, we may clearly recognize other classes which differ, both as to their origin and composition. For example, we may readily distinguish from all others the deep red soil of the argillaceous slates from that of gneiss or granite, though the latter has a deep red color also, or, from the deep red soil of the sandstone of Orange, Chatham, Moore and Anson. There is also another peculiar soil which skirts the northern counties, Granville, Person, Caswell and Rockingham. It is adapted to the growth of fine tobacco. It is a light gray soil.

The soils, however, which form the subject of this report, occupy the eastern counties of the State, and may all be regarded as marine products with one exception, the vegetable soils, which occupy the swamps and pocosins of the extreme eastern part of the State. The others which have been referred to are derived immediately from the rocks upon which they rest, and have been formed by atmospheric agencies.

The vegetable soils, on the other hand, were formed by the growth of vegetables which have long since ceased to live, and have undergone disintegration in a greater or less degree; some are coarse and fibrous, others exist as a close compact mass of vegetable matter, perfectly disorganized and in the best condition possible for cultivation. The mass remains *in situ*, frequently homogeneous, and may be cut into blocks and dried like brick.

I have applied to these vegetable accumulations the usual term soil, for the reason that they are cultivated and frequently productive. Others probably come more properly under the common name peat, as the mixed earthy matter is too small to be cultivated without the addition of earthy matter, and have remained *in situ*, and undisturbed since their seeds took root.

The peculiarity of this vegetable soil then consists in its composition, and the interest which is especially attached to it arises from the small amount of earthy matter which it contains. It gives us, therefore, an opportunity to determine the smallest amount of earthy matter compatible with remunerating crops. It is also proved by observation that all crops require earthy matter,—it may be comparatively small, but if the inorganic matter is reduced to a certain small percentage, the crop fails, although it is placed, in one sense, in a magazine of food. The determination of the smallest percentage of inorganic matter which is compatible with a good crop, is practically important. Large tracts of land in North Carolina consist of organic matter, with too little soil to permit of its cultivation. If inorganic matter is added, it will make it productive, and possibly valuable. But how little is required, how much expense may be required to bring it to or put it in a cultivable state, is a legitimate inquiry, and one which may be productive of considerable profit. It is evident, however, that in a country like this, where there are vast areas of wild land to be subdued, that these lands under consideration cannot come in competition with good soil at government prices, unless it can be shown that the expense of reclaiming them is comparatively small; still, the question sought to be determined is an interesting one, and I have attempted its solution, the results of which will be given in the subsequent pages.

§ 7. A secondary fact requires a passing notice. While all the elements enumerated are essential to a good soil, some are more so than others. Thus, certain plants require potash, while to others this element is not so essential, or it holds only a subordinate place. In wheat it is very necessary, while to clover it is less so, and in the latter lime seems to take its place. As a general law the most expensive elements, as potash and phosphoric acid, abound in the seed and fruit, while lime is most usually found in the wood and bark or stem.

Silex in the cereals is an essential element in the stem or stalk. Its office is to give it strength and hardness.

Each element, therefore, being destined for a particular organ, performs or fulfils a certain office or function.

These specializations we may regard as predetermined results, effected through the instrumentality of the cell force; but how, it is impossible to say; how the salts or compounds of phosphoric acid are carried up to form the seed and there remain and accumulate, and how the silex is arrested and accumulates in the stem, it is impossible to say.

We may be assured, however, that the machinery of a plant will work right if it is fed with the necessary food. Knowing, therefore, what a plant wants, it becomes the special business of the farmer to supply it. The perfection in agriculture will consist in a strict application of the doctrine of specialities, and this specialization will not be confined to a supply of food simply, but will extend to the mechanical cultivation: each plant will no doubt be found to do or grow better under a certain mode of cultivation.

§ 8. Sandy soils predominate to a great extent over all others in the eastern counties, though there are tracts in which clay is in as great excess as sand. The extreme varieties may be summed up as follows: 1st, sandy soil to an excess which destroys cohesion and becomes blowing sand; 2d, clay; 3d, vegetable soils to such an extent as to exclude earthy matter, or to contain merely some 4 or 5 per cent. of it.

Between the extremes, as enumerated, there exist mixtures in various proportions, as usual, except that, as a general rule, the proportion of sand is somewhat greater than in the soils belonging to other parts of the State.

As an example of soil in which sand is in greater excess, I may state that the following is an instance worthy of note. The specimen was taken from Bladen county, near Elizabethtown, and represents a kind common to that section. Thus,

Silex,	94.80
Water,	1.20
Organic Matter,	1.50
Per oxide of iron and alumina,	65
Lime,	01
Magneisa,	trace,
Potash and soda,	traces.

The essential constituents of a good soil in this example exist only in the smallest proportions,—and though it produces plants, yet the valuable elements exist in too small proportions to pay for tillage.

The great excess of sand is, however, palpable, and it is also evident that there is a great deficiency of clay or alumina, which gives consistency to soils, and which forms the basis upon which fertilizers may be profitably applied.

It belongs, it will be conceded, to a particular class, as there is a single element in great excess. Although there is a great excess of sand in these examples, to which many more might be added, still, this excess, in itself considered, does not disqualify them for the growth of certain crops, particularly the ground pea, though it is possible their constitution may not be fully adapted to that crop, yet so far as the proportion of sand is concerned it is not in excess. This fact is stated for the purpose of alluding to what may not be known to many, that a soil which is really poor and unsuitable for one crop, may be well suited to another. The quality of the crop may be much better when grown upon a soil where sand is in great excess than upon a rich and well proportioned soil.

§ 9. The contrast between soils, one of which is not well proportioned, while the other is, is strikingly exemplified in the composition of another soil from Halifax county. Thus, I found:

Silex,	74.80
Water,	21.90
Organic matter,	5.40
Alumina and per oxide of iron,	14.00
Phosphoric acid,	01
Lime,	40
Magneisa,	20
Potash,	05
Soda,	03

Another from Halifax county resembles very closely the former; thus, I found on submitting it to analysis:

Silex,	94.15
Water,	1.30
Organic matter,	1.35
Oxide iron and alumina,	1.80
Lime,	15
Magneisa,	01
Potash,	01
Soda,	01

Another soil from Halifax which had been long under cultivation, but whose composition is somewhat better; thus, it contained:

Silex,	92.56
Water,	1.20
Organic matter,	2.70
Oxide iron and alumina,	2.70
Lime,	13

Magneisa,	24
Soluble Silica,	10
Potash,	trace,
Soda,	18

The presence of phosphoric acid was not determined in either of the foregoing, but as it is in combination with the small per centages of oxide of iron and alumina, it is evident that it exists in proportions less than that of the alkalies.

The soils of Halifax, were originally sandy, yet the relative proportion of sand, as they are now constituted, is considerably greater than when they were first brought under cultivation. The soluble matters, those consumed by the crops which they have borne, having been removed with them, and nothing returned to supply their places, they are yet capable of bearing very light crops, but it is doubtful whether the cultivation of land so poor as these really pays. If an example of poor soil is placed side by side with a good one, the comparison is much facilitated:

	GOOD SOIL.	POOR SOIL.
Silex,	74.80	94.15
Water,	4.90	1.30
Organic matter,	5.40	1.35
Alumina and per oxide of iron,	14.00	1.80
Phosphoric acid,	51	
Lime,	40	15
Magnosia,	20	01
Potash,	25	01
Soda,	13	01

In making a safe comparison between the composition of good and poor soils, it should be stated that less alumina and iron would not displace the soil from the position I have placed it. The silex is in the proper proportion, and the organic matter may be regarded also as sufficient, though as we shall see in the sequel, this element may be greatly increased to the advantage of long cultivation. Where it is wholly absent, seed fails to ripen; a fact which shows the necessity of its presence. Silex is the basis of all soils, and where it is entirely absent, barrenness is certain. It is soluble under needful conditions, and it enters largely into the straw of all cereals.

Alumina never enters into the composition of plants at all; but it performs an important function notwithstanding; it holds as it were the particles of earth together. Its true office may undoubtedly be shown by experiment. Pour water upon a soil well charged with clay, and it remains upon the surface; but poured upon sand, it quickly disappears. If the water was charged with fertilizing matter, this also will remain, and be held near the surface by the clay, and within reach of the roots of the plant.

§ 10. The fact is well known that sandy soils do not retain manures; while on the contrary, clay soils retain all fertilizing matters with great force. Clay indeed absorbs ammonia under all circumstances, and it cannot be entirely dissipated or driven off short of a red heat. It obstinately retains water. Some of the functions of clay are performed by other elements. Lime and iron and organic matter, for example, give cohesion to soils, and aid in the retention of water.

Water exists in soils in two conditions. In the first, it seems to adhere to the surfaces of particles, and hence is liable to constant variation. This is hygrometric water. In the second, it forms a constituent part of the salts in the soil, as the soluble salts of lime and alkalies, the crenates, etc. In the first instance, it is mostly dissipated by an exposure of 400 degrees of Fah., while a heat near to redness is required to remove it from the organic salts.

All the elements which have been enumerated, except alumina, enter into the constitution of plants; but as I have had occasion to say, in different proportions in different plants, and also in different proportions in the parts of plants.

An example or two of soils occupying another extreme, where the organic matter is in great excess, may be cited from localities in Tyrrel and Carteret counties. In the former county, large tracts lying upon Croatan Sound, furnish organic matter in great excess, and at the same time they are deficient in the earths. Thus in an uncultivated soil I found it composed of

Organic matter,	92.70
Sand,	6.02
Lime,	0.02
Phosphate of lime, alumina and iron,	0.90
Potash,	0.20
Soda,	0.06
Magnesia,	trace.

The silex in this case is a white marine sand which becomes visible after rains, or after a year or two of cultivation. It is too coarse to furnish the necessary amount of soluble silica for a succession of crops. When the vegetable matter is removed, it remains as a white sand still, and is blown into ridges.

§ 11. The condition of the vegetable matters, as in the case of the other elements, is quite variable. Sometimes it is very fine, and is thoroughly incorporated with them; in other instances it is coarse, or in the condition of fibres. In the former state the sand is not so readily exposed; in the latter it is always visible, and is indicative of a poor condition, or of its unsuitableness for cultivation. It has not been exposed long enough to change it to the condition required for crops of the most valuable kind.

A still more remarkable case of excess of vegetable matter composes a tract in Carteret county, and is known as the open ground prairie. This tract, or that portion of it lying within a certain zone of rich and productive land, contains a growth of sphagnum or moss, together with other vegetables intermixed, with which there is only a minute quantity of earth. I obtained it from a depth of 18 inches, and it gave only 3 per cent. of inorganic matter, and this was mostly the ash of the vegetable fibre. This case furnishes an example of an unproductive soil, so far as the grains are concerned. The outer rim of the open grounds is an excellent soil.

Much has been said respecting the open ground prairie, and enquiries are now frequently made respecting the character of this tract; and whether it is susceptible of a profitable cultivation. As the soil is now constituted, a kernel of corn planted in it would germinate and grow well apparently until it is about one foot high, when it turns yellow and dies. It is then evidently in an uncultivated condition.

The question then comes up, can the open prairie be made cultivable artificially, and if so, how? The question first put is not designed to inquire strictly into the possibility of the thing, because all who have given some thought to the question, know very well that it is *possible*, because a soil can be made from the start, by putting together the proper elements. and this can be done with the open ground prairie; but can it be done profitably? Now, when we are assured that the soil of the open prairie ground is composed exclusively of vegetable matter, it is plain, that the earths must be added to give it the composition required for the perfection of vegetables of any value to man. The old practice consisted mainly, in giving peaty soils (as this must be ranked in that class,) a heavy dressing of lime. It is evident on reflection, if the principles in the foregoing paragraphs are correct, that this practice could not be relied upon, for it would only acquire a single element. Something more is wanted. Not only lime, but iron, alumina and silica are required. We may infer that the phosphates and alkalies will be supplied by the decay of vegetable matter, and, from this fact, it appears at least plausible, that the treatment which the open ground prairie demands, is the addition of some natural soil. It may be taken from the nearest marsh where mud or soil may be obtained, provided it contains silex, alumina, iron, etc.

Knowing, then, what substances are wanting in this soil, and hence what must be added, the question resolves itself into this: how much does a soil of the description of that under consideration require to make it productive? We have seen that the soil upon Croatan sound is at least tolerably productive, which contains only 7.30 per cent. of inorganic matter, and that the element which greatly predominates over the rest, is sand, in a state unfitted to furnish soluble silica. We may regard the Croatan soil as containing the smallest quantity of earthy matter, and at the same time possessing the ability to grow the cereals. Leaving the sand out of view, we may infer that the least quantity of earth which is required to the open ground prairie will be not less than 140 to 150 tons to the acre. When this expense is added to the expense of drainage, it is evident that in a country where land is cheap it would not be economical to expend so much money and labor to create as it were a soil adapted to the better class of vegetables.

§ 12. The effect of cultivation of soils composed mainly of vegetable matter and marine sand, is to consume so much of the former that the latter becomes in its turn predominant, and even after a few years' cultivation only, the white sand shows itself through and upon the surface of the black vegetable matter, and soon afterwards it appears in sufficient quantities to form white ridges over the cultivated field. When this takes place, the soil has already begun to exhibit unmistakable evidences of partial exhaustion.

The soils in which vegetable matter predominates, apperently in great excess, not injuriously however, prevail over large tracts or areas in the eastern counties, and are beginning to be esteemed the most valuable lands of any in North-Carolina. They are not confined to one or two counties, but may be found in most of them which lie east of the Wilmington railroad. They also prevail in the south-eastern section, especially in New Hanover and Columbus.

Some of the tracts are classed as pocosin and swamp lands, but they agree in having a very large percentage of vegetable matter, and in being also thoroughly wet and frequently covered with water, I have found that there is no constant percentage of vegetable matter where different and distant tracts are compound together. It is as variable as the clay or sand in argillaceous and sandy soils. There is also a variableness as to its condition; it is often perfectly disorganized and presents a compact appearance when cut into blocks; or it may be in the condition of coarse fibres with their texture or structure perfectly preserved. In the first case, it is in the proper condition for cultivation, and the latter, it has not passed into that state and condition which is fitted for the nutrition of the cereals. The coarse vegetable fibre predominates in the open prairie grounds of Carteret, and the former in those of Hyde and Onslow counties. So also these vegetable soils vary endlessly with respect to the amount of soil and sand. The Hyde county soils may be regarded as the standard soils for excellence of this class, and hence it is important to determine their composition. On their own account, it is important to determine the composition, as well as for the purpose of comparing their composition with others which resemble them in their external characters. Many mistakes have been made in the swamp lands; for when wet and examined in the ordinary way they look rich—with the presence of a superabundance of vegetable matter, their true characters may be concealed. In many cases the condition of the earthy matter is overlooked. It may indeed be too small; or it may be a coarsish marine sand destitute of fine earth. In all cases it is possible, and indeed easy to determine whether it will be productive or comparatively valuable. This is an important fact to make out, for all these lands require to be drained thoroughly, and it is certainly an object worth attention to be able to determine before hand *whether the tract is worth the expenditure before it is incurred.*

The Hyde county soils have acquired a deservedly high reputation for fertility. Some tracts have been cultivated over a century, and the crops appear to be equally as good as they were at an early period of their culture; and yet no manure has been employed, and they have been under culture in indian corn every year; or what would be equivalent thereto. If this crop has been omitted, wheat has been substituted for it; not because they are properly wheat soils, but if they are uncultivated, the weeds acquire a size that it is impossible to cover them the next year. The same difficulty occurs in part in the culture of corn; the stalks are so numerous and large that it is difficult to bury them so completely that they shall be concealed, and preserve at the same time an even handsome surface. For this reason critics of a morbid stamp have said, that the Hyde county planters are slovenly, overlooking the facts refered to, which are really the sole causes of the defects complained of. Though the defects are not very palpable under any circumstances, still it is sometimes useful to a community to have faultfinders, and to have their doings overhauled by a would be *wise critic.*

§ 13. Hyde county appears to be nearly a dead level. It rises of course a few feet above the sound, but it is imperceptible to the eye. Buildings may be seen for great distances, and were the whole surface laid out in proper order, it might be made to appear like an immense park. The depressions of the surface are due to fires which have consumed the vegetable matters to the depth of from four to ten and perhaps fifteen feet. In these depressions the surface water has accumulated, and in a few instances large lakes are the result. Mattamuskeet lake is the largest of the surface drainage. Its former extent was not less than twenty miles. Its circumference now exceeds sixty miles by the road,—and as the traveller proceeds on his route, there is nothing more surprising than the succession of corn fields which are always in view.

The most common natural growth of the best swamp land of Hyde county is cypress and black gum.

In one respect this region differs from others farther from the sea. There is no difficulty in the cultivation of the grasses. It is evident the climate is more humid, and the sea breezes moderate the heat sufficiently in summer to favor the developement of this family of plants. There is no doubt, also, that if the attention of the planters was turned to the cultivation of grasses adapted to the climate, greater profits might be realized than from the cultivation of maize. It is less expensive, and as hay bears a high price, and is obtained from a distance, in all the villages of this part of the State, and as there is always a communication with them by water, there can be no doubt that the profits which would arise from hay making, would considerably exceed those of corn. The green surface of the lake shore, the yards of the houses, and the appearance of the small pasturages sustain this view.

§ 14. The peculiarities of the soil of Hyde county, that particularly of the lake region, are comprised in two particulars: 1st, the large quantity of fine vegetable matter they contain; 2d, the extreme fineness of the intermixed earthy matter. The earthy matter is invisible in consequence of its fineness, and is evenly distributed through the mass. An inspection of it even under a common lens will deceive most persons, and they would be led to infer that it was entirely absent. Unlike other soils it contains no coarse visible particles of sand; and hence it appears that during the growth of the vegitables which form at least one-half of the soil, it was subjected to frequent overflows of muddy water; or else the area over which these peculiar soils prevail was usually a miry swamp which communicated with streams which brought over it the finest sediment of some distant region. This sediment is frequently a fine grit, and fine enough for hones, and when the vegetable matter is burnt off, it assumes a light drab color. The character of the Hyde county soils has never been understood. The cause of their fertility has never been explained, and many persons who are good judges of land have overated the value of swamp lands in consequence of the close external resemblance they have borne to those of Hyde. Analysis, however, will in every case detect the difference between the common swamp soils, and those of Matamuskeet lake.

It is unnecessary to dwell farther upon the points I have stated respecting the characteristics of these remarkable soils. It will appear in the sequel that there is a great uniformity in the composition of these soils, both as it regards the amount and condition of the vegetable matter, and the quantity and condition of the fine grit intermixed with it.

Regarding as I do these soils as the proper standard for the valuable swamp soils of the eastern section of the State, I have subjected many samples to a rigid chemical analysis.

The result of these analyses have thrown much light over them, and explains satisfactorily their steady productiveness for long periods. It will appear that their fertility is due not only to their vegetable matter, but also to the composition and condition of the earth in combination with it.

Hereafter, it appears to me, it will be unnecessary to subject soils of this character to a strict analysis, for reasons which will be stated in the sequel.

In my journey to Hyde my principal objects were to select the standard soils for analysis, and to investigate upon the ground, the peculiar conditions which seemed to favor the production of indian corn; for of all crops this seems to be the one to which the soils are specifically fitted.

In accomplishing the objects of my visit I was ably seconded by Dr. Long, of Lake Landing, who has become the owner of a tract which has borne this crop for one hundred years without manures. It does not seem to have deteriorated by this long cultivation; or the crops do not *show* a perceptible falling off; still there has been a large consumption of materials during the one hundred years of cultivation which may be made to appear by analysis. The great supply of nutriment, however, still holds out, and the one hundred years to come, if subjected to no greater drains upon its magazine of food, will, at such a distant period, continue to produce its ten or twelve barrels of corn to the acre.

CHAPTER II.

The best soil of Dr. Long, of Hyde county—its composition—its common yield per acre of corn. Mr. Burrough's soil of the north side of Mattamuskeet Lake. Amount of inorganic matter which a crop of corn removes from the soil. Each organ to be furnished with appropriate nutriment. Maize an exhausting crop. Soils from the plantation of Gen. Blount, Beaufort county. Gen. Blount's letter, etc.

§ 15. The soil which Dr. Long regarded as his best, and which had been under cultivation only three years, I shall now speak of, and state its composition, and present it as representing very nearly the original condition of the best soil of the county. It is rather light and loose, of a black color like all vegetable soils. It is not however spongy. Rains do not expose grains of quartz as in many instances of the gall-berry lands. It becomes rather lumpy on drying. Its composition is as follows:

Organic matter,	48.10
Silex,	43.00
Oxide of iron and alumina,	6.40
Lime,	0.21
Magnesia,	0 12
Potash,	0.16
Soda,	0.18
Chlorine,	trace,
Soluble Silex,	0.03
Sulphuric acid,	0.04
Phosphoric acid,	0.30
	98.55

The silex, after the removal of the organic matter, is of a light drab color, exceedingly fine, or nearly fine enough for sharpening fine edge tools. If all the vegetable matter was removed, this fine earth would probably be too compact and close for cultivation; but, intermixed as it is with the debris of vegetables, it is sufficiently porous to admit all the light and air required for the luxuriant growth of any crop which may be put upon it.

The composition of this soil, it is evident, shows a large proportion of vegetable matter. This is intimately blended with fine earthy matter, the basis of which is silex. In combination with it we find a full proportion of iron and alumina, or clay, which gives coherency to the grains, and besides the nutritive elements, lime, magnesia, potash, phosphoric acid, exist in as large proportions as in other rich and productive soils. The regular yield of this soil to the acre is from ten to twelve barrels of Indian corn. In favorable seasons it amounts to twelve, in less favorable it may reach only ten barrels. It is also easy to cultivate.

The composition of a soil of a similar character, and which has been under culture by Mr. Burroughs, of the north side of the lake, is as follows:

Silex,	34.60
Water,	12.30
Organic matter,	41 90
Peroxide of iron,	3.70
Alumina,	5.10
Soluble silica,	0.40
Lime,	0.48
Magnesia,	0.27
Potash,	0.13
Soda,	0.10
Phosphoric acid,	0.12

This soil, though exposed in paper in a dry room for two months to the air, contained more water than the preceding. Its composition should be calculated without the water. So it is probable that the phosphoric acid, if obtained and calculated from the full proportion of earthy matter, would show a more striking result. But it is evident that there can be no deficiency of this important element, inasmuch as the crop is one which is necessarily rich in phosphates. The depth of this rich vegetable soil varies from 5 to 10 feet, rarely less than five feet. This may be taken too as the usual depth of the soils of this description, not only in Hyde, but in all the eastern counties where swamp and pocosin lands prevail.

§ 16. There are but few instances on record, where a soil has been under cultivation a century, and still retains its apparent original fertility. It must of course have lost a large amount of phosphoric acid, potash and lime; still the crops are equal in measure to what they were when first cultivated. In order to test the value of a soil which had borne a crop for one hundred years, and during the whole period had not received a bushel of manure, I selected a parcel of it at a distance from buildings, or from a spot which could not have received any artificial aid.

This parcel gave the following result, on submitting it to analysis:

Silex,	59.00
Organic matter,	22.20
Peroxide of iron and alumina,	8.00
Lime,	0.10
Magnesia,	0.09
Potash,	0.02

Soda,	0.03
Sol. silica,	0.20
Water,	8.90
Phosphoric acid,	trace.
	99.44

These remarks are justified on comparing the results of this analysis with Dr. Long's soil, which has been under cultivation only three years; thus, the silica is in greater proportion, and the organic matter, less; and it is due no doubt to the fact that it has been under cultivation for the time specified. It still retains, however, a magazine of food for future crops; and if not exhausted at a greater rate than during the last century, it will be a rich soil at the close of the next century. It will be perceived that all the elements of fertility which belong to new and unexhausted soils still belong to this. The inorganic matter is extremely fine, like the finest grit, and in the proportion required for the production of the most valuable crops. Growing, as we perceive, in a magazine of food, it seems to show that it is a crop upon which it is scarcely possible to overmanure, and that it is unlike other corn crops, which, when over supplied with food, run to stalks and leaves to the detriment of the grain.

§ 17. If we calculate the amount of inorganic matter which a hundred crops of maize remove from the soil, we should find it to amount to many thousand pounds.

From data in my possession, I am led to believe that five hundred pounds per acre of inorganic matter is removed in every crop. This inorganic matter is contained in the kernels, cobs, husks, silks, leaves, sheaths, stalks and tassels; each organ containing its own appropriate amount.

The number of plants which are allowed to grow upon an acre, amount to fourteen thousand and seven hundred. Each plant removes from the soil a specific amount of the earthy compounds, and nearly in the following proportions, viz:

In Silica,	195	lbs.
Earthy phosphates,	108	"
Lime,	25	"
Magnesia,	18	"

Potash,	78 "
Soda,	30 "
Chlorine,	29 "
Sulphuric acid,	34 "
	507

If five hundred pounds of the earthy constituents of this soil are removed from one acre in one year or in a single crop, it will amount in one hundred years to fifty thousand pounds—a quantity which would exhaust most perfectly any of the ordinary soils of the country.

In an analysis which I have made of the kernels and cobs of the yellow corn, I found:

	COBS.	KERNELS.
Silica,	4.67	5.93
Earthy phosphates,	8.22	22.18
Lime,	0.10	0.10
Magneisa,	0 30	1.50
Potash,	12.31	14.95
Soda,	2.03	14.11
Chlorine,	0.04	0.39
Sulphuric acid,	0.11	2.74
		61.81

That the composition of the leaves may be compared with the foregoing, I subjoin an analysis of the leaves made at the same time and growing upon the same plant:

	LEAVES.
Silica,	82.88
Earthy phosphates,	29.27
Lime,	9.40
Magneisa,	1.91
Potash,	19.70
Soda,	13.14
Chlorine,	15.07
Sulphuric acid,	6.46

It might be supposed that as the sheaths of the leaves belong in one sense to the leaves themselves, that their composition would be the same; but this is not the case as may be seen by the following analysis:

	SHEATHS.
Silica,	39.66
Earthy phosphates,	7.54
Lime,	1.58
Magnesia,	58
Potash,	5.57
Soda,	9.26
Chlorine,	2.20
Sulphuric acid,	8.92

In the sheaths the earthy phosphates and alkalies are much less than in the leaves. In the cobs too the earthy phosphates are less than in the kernels; it seems, therefore, that each part or organ has its own peculiar composition. To complete this view of the composition of the plant of the maize, I subjoin an analysis of the stalks; thus, they contain:

Silica,	8.78
Earthy phosphate,	10.30
Lime,	1.92
Magnesia,	0.64
Potash,	11.08
Soda,	17.09
Chlorine,	7.42
Sulphuric acid,	7.38

It should be observed that these several analyses were made of a single plant, and the proportions are those belonging to the plant, or its parts, and not properly percentages. The ash was obtained from all the leaves, or stalks, and kernels, and the whole ash obtained analyzed. Hence the difference of composition of those parts are presented in a strong light, as well as in a true proportion.

From the foregoing it will be perceived that where a crop is to be manured or a fertilizer applied, it is not sufficient to apply the earthy phosphates, for we perceive that every organ or part requires all the elements which we find in them. The notion, therefore, should be dispelled, that bone earth is the main fertilizer for the maize crop, or that it is enough to furnish substances which consist of elements found in the grain or fruit. For the perfection of the crop it is necessary that the leaves and stalks, tassel and cobs should be furnished with appropriate elements of food as well as the grain; for that the grain may ripen and acquire perfection, the leaves and stalks also should be equally perfected. It can scarcely be doubted that the grain itself depends for its full development upon the perfection of the parts which precede it. They are the organs which bring up the nutriment from the soil. Remove the leaves at an early day, and the grain is destroyed, or never comes to maturity; but supply matter suitable for their increase and perfection, and the grain is supplied also. It will be observed that the different subordinate parts frequently contain elements which are not found, except in very small proportions, in the seed or grain; yet, there is no doubt these elements are quite essential to the perfection of the plant.

§ 18. Maize must be ranked among the most exhausting crops; and it is evident that poor soils will scarcely repay the farmer for its cultivation. It is evident that, unlike other cereals, there is little danger of using too much manure in its cultivation, as it will bear almost any amount without injury, provided all the elements of fertility exist in the magazine of food provided for it. It is not liable to run to foilage, and thereby fail to produce grain; neither will it lodge or fall down by its own excessive disproportion of organic to its inorganic nutriment.

While it must be admitted that maize is an exhausting crop, it is equally clear and conclusive that it is one of the most important and valuable, and hence it may be regarded as one which pays the best.

§ 19. The foregoing remarks respecting the maize crop have been made in consequence of the peculiar adaptation of the soil of Hyde county to this cereal. It is the granary of the South. It is true that the number of bushels per acre which constitute the average crop is less than the number frequently made on other kinds of soil. Thus a hundred bushels of corn may be grown upon an acre, but the Hyde county soils rarely exceed sixty bushels per acre,—but from fifty to sixty bushels are grown annually per acre for an indefinite term of years, without the expense of fertilizers, while the heavy premium crops require a great expenditure on them; and these have to be repeated in order to keep the ground in a good condition; and hence, in the long term of years, the profits of these rich lands greatly exceed those which are only moderately so, naturally, and require every few years an instalment of manure.

§ 20. The similarity in the composition of the soils and lands surrounding Matamuskeet lake in Hyde county is remarkable. They are all eminently rich in vegetable matter, and all are supplied with a sufficiency of fine earthy matter; in which respect they differ greatly, as will be perceived from the open ground prairie in Carteret county. The similarity appeared so great that I have not multiplied analyses of them. I have, however, specimens received from Gen. Blount, from Beaufort county, which I have analyzed; all of which will go to show that there is an extension of similar swamp lands of that direction in the county of Beaufort, which I have submitted to analysis; all of which go to prove the extension of the Matamuskeet lands westward, or of swamp lands quite similar in composition to these justly celebrated soils.

The soils which were collected by Gen. Blount were four in number, and were taken from tracts, some of which had been under cultivation several years, while others were comparatively new.

After having submitted these soils to analysis, I stated to Gen. Blount my opinion of the samples I had operated upon, and requested a statement from him also of all the facts connected with them which he regarded as of sufficient importance to be made public.

In reply to this request I received the following interesting communication which I propose to incorporate with this report.

It should be stated, however, for the benefit of those who are not acquainted with Gen. Blount's husbandary, that he has been engaged in the successful culture of swamp lands between forty and fifty years, and hence is amply qualified to express an opinion respecting their productiveness and value.

The following is the communication referred to:

MADISONVILLE, (NEAR WASHINGTON,) BEAUFORT COUNTY,
January 30*th*, 1858.

PROF. EMMONS—*My Dear Sir:*—Your letter was duly received. I will now give you a description of the land of which the four parcels sent you are specimens:

No. 1.—A dark soil, from fifteen to twenty inches deep, incumbent on porous clay, with some fine sand intermixed; through this substratum the water percolates freely. The natural growth on this land, (before being cultivated,) was a heavy growth of black gum, a scattering growth of large poplars, some maples, a few laurels; here and there a large short strawed pine. This land has been cultivated in corn for three years, and has produced from 40 to 50 bushels per acre.

No. 3.—When cleared, some ten years since, was considered by me second quality swamp land. The growth is formed of gums, but more laurels, pines, and poplars than No. 1. For ten consecutive years it has been cultivated in indian corn; when in its prime it produced 40 bushels per acre—the last crop 30—the past season it was sown in oats, produced 20 bushels per acre. The specimen sent you was taken from the poorest spot I could find in the field, (judging from the growth of oats then on it;) the soil where the specimen was taken from was about 12 inches deep, the balance of the field 18.

No. 2.—Unreclaimed swamp—soil from 18 to 24 inches deep; subsoil a different clay from that which underlays the previously described land, it is lumpy and resists the spade. My opinion is that the water does not pass freely through this subsoil, and consequently the surface soil is wetter than on the lands above mentioned. The natural growth of this land is: reeds standing very thick, of moderate size, small sickly pine saplings, red and white, bay bushes and gallberry. I have no doubt that this land has been often burnt. I find strata of ashes at different depths below the surface, and the stumps of large pine trees charred. I own about 3000 acres of this description of land—it lays between the long leaf pine land and the gum lands, and is the greater part of the year filled with water to the surface. For some time after every heavy rain the surface is partially covered, and the water slowly disappears; every foot of it can be drained; it adjoins my farm. Why should not such land, when thoroughly drained, be fertile? If it would not be, what should be the proper treatment to make it productive?

No. 3 lies between Nos. 1 and 2.

No. 4.—Soil of the complexion of the specimen sent you. It is from 2 to 3 feet deep; incumbent on soapy clay, which is porous, and allows an easy descent of the water. The growth of timber on this land is magnificent: black gums, from one to two feet diameter at the stump, fifty to sixty feet to the limbs, straight bodies, the limbs not drooping, but forming with the body an angle of about 30 degrees, limbs and twigs showing that the growth is healthy and vigorous; a few very large, long bodied poplars; some maples, corresponding in appearance, as regards size, &c., with the gums above described; cypress trees, averaging from 8 to 10 in number per acre, from two and a half to four and a half feet diameter at the stump; one hundred feet to the limbs, straight bodies, small bulky tops, limbs not drooping but erect. I have none of this land in cultivation, but have just commenced to reclaim it. My opinion is it will be found equal in production to the lands on the south-side of Matamuskeet lake.

On a farm laying on said lake that I once owned I have made one hundred and fifteen bushels of indian corn per acre, and thirty bushels of wheat per acre. I think this last described land, No. 4, with perfect drainage and judicious cultivation, will produce as much as the Matamuskeet lake land spoken of; appearances, however, may be deceptive.

I have been, for a period of forty years, engaged in reclaiming and cultivating swamp lands, such as I have described, and have found it a profitable business. I am located near the margin of the swamp, (of which my plantation is a part;) it contains about 30,000 acres, and is south of my residence. The health of my family, white and black, will compare favorably with the healthiest locations in eastern North-Carolina.

We have, as you are aware, large bodies of rich swamp lands in this portion of the State. Within a few years wealth and population has flowed, and is still flowing in upon them, which promises the happiest results to the good old North State. Rich swamp land, like almost every thing else, will show after a while the effects of bad treatment, but fortunately for us, if we impoverish our land by severe and injudicious cultivation, we have in close contiguity inexhaustible supplies of shell marle, which has proved itself a panacea to worn down swamp land. Guano and the other manures in common use produce as fine, perhaps a better effect, on swamp land than any other description of land of which I have any knowledge. I fear, sir, I have taxed you too severly; the interest I feel as a citizen of the eastern part of the State I mention as my justification. Should you wish more specific information than I have given, it will afford me pleasure to furnish it.

Such is my great aversion to writing, I have been compelled to enlist the aid of my daughter, Mrs. B., who is now with me. You will perceive that a lady has been my amanuensis.

Most respectfully,

WILL. A. BLOUNT, SEN'R.

From the foregoing communication the reader will be prepared to form a correct opinion of the character of the swamp lands referred to, especially when taken in connexion with their composition as determined by analysis.

No. 1.—On being exposed for a few weeks to the air becomes dry. Its color is blackish brown, it contains undecomposed bark, wood and some roots, but is mostly made up of decomposed vegetable matter. The earthy part is not visible as in many vegetable soils of the poorer class.

On submitting it to analysis I found it composed of the following elements:

Silex,	65.540
Organic matter,	26.100
Water,	6.050
Peroxide of iron and alumina,	4.920
Carb. lime,	0.490
Magneisa,	0.050
Potash,	0.003
Soda,	0.020
Phosphoric acid,	0.003

The silex, as in most of the good swamp soils, is extremely fine. Its color is drab, and hence probably contaius a small quantity of alumina which cannot be detached without being attached by potash.

This soil, it is evident, still contains the elements of fertility, and it is also evident that it will bear cultivation for years to come without exhaustion. It will be observed that the natural growth upon this soil is one which indicates fertility, as the poplar and black gum, and a large growth of short leaved pine, the growth being very heavy.

No. 2.—This specimen or mass of soil consists apparently of vegetable matter without any earth. It is black, and preserves a moist state, though it has been exposed to the air in a box for several months; and on being exposed in a drying oven lost its moisture very slowly. It contains fresh vegetable fibres, portions of partially decomposed wood and bark, etc. Still it is rather homogeneous, and is unlike the coarse fibrous soil of the open prairie of Carteret.

On submitting this soil to analysis, I found it composed of the following elements:

Silex,	74.600	74.600
Organic Matter,	18.000	18.100
Peroxide of iron and alumina,	3.100	3.100
Phosphoric acid,	0.021	trace,

Lime,	0.049	0.040
Magnesia,	0.005	0.005
Potash,	0.040	trace,
Soda,	0.030	trace,
Water,	4.000	4.000
	98.845	99.845

This soil was dried before the quantity was weighed for analysis. When exposed to about 300 degrees of Fah., it lost fifteen per cent. of water.

This soil has not been cultivated, and though it looks rich, still I am inclined to regard it as a poorer soil than No. 1. It contains more sand, is rather coarser, and less alumina, iron and vegetable matter. The alkaline earths, as lime and magnesia, are much less. The same may be said of the alkalies, potash and soda. The depth of this soil is from eighteen to twenty-four inches, resting on a hard and rather impervious bottom. Its natural growth is also different; as it consists of reeds standing very thick, and small sickly pine saplings, red and white bay bushes, gallberry, etc.

This growth, it is evident, might be due to the impervious bottom, or its low temperature; but it is also in part due to the absence of the most important elements of fertility. There is no doubt, however, but a low temperature, which is due to the presence of water, is competent to produce an apparent sterility, low bushes of peculiar kinds, as bay, gallberry, alder and willow.

No. 3.—The color of this soil is a dark ash or gray. It has become dry in the box in which it was sent, while No. 2 has remained wet. It is pulverulent and light, though somewhat lumpy. The vegetable matter exists evidently in a large proportion, yet a close observer would perceive that it is less than in No. 2.

On submitting it to analysis, I found its composition as follows:

Silex,	81.600
Vegetable matter,	12.800
Peroxide of iron and alumina,	4.100
Carb. of lime,	0.020

Magnesia,	0.010
Phosphoric acid,	trace,
Potash,	trace.

This soil was regarded by Gen. Blount as second quality. Its growth consisted of low pines, gums and poplars. It however produced forty bushels of corn to the acre, but the last crop was only thirty bushels. Afterwards, it gave twenty bushels of oats to the acre.

The proportion of silex, it will be perceived, is much greater than in No. 1. The specimen was taken from a poor spot in the field. It had been under culture for ten years. Depth of soil twelve inches.

In attempting the solution of the question, why a poor crop was at last produced, we should not forget that certain soils in this climate become dry at an early day; and if so, we invariably find the cereals growing very slim and slender, and perhaps soon cease to grow, turn yellow, and produce, if any, a very small ear of grain. In a shallow soil such a result may be expected, notwithstanding the soil, on analysis, may be found to contain the elements of fertility. In the same field, plants growing in the same soil, a part may yield seed and fruit, and another will fail; the results being dependent on the existence of moisture surrounding the roots of the plant.

No. 4.—The color is grayish black, and contain half decomposed roots, bark, etc. It has also partially dried in the box, and in drying, becomes lighter colored. This soil is deeper than either of the preceding, being between three and four feet deep, and incumbent on a porous bottom.

The growth is very large, consisting of black gum from one to two feet in diameter, and from fifty to sixty feet high. The limbs are straight as well as the bodies. Very large poplars also are found scattered over the field, also cypress in clusters from eight to ten in each.

This sample I found composed as follows:

Silex,	77.500
Organic matter,	15.400
Peroxide of iron and alumina,	6.900

Lime,	0.500
Magnesia,	0 100
Potash,	0.019
Soda,	0.029
Phosphoric acid,	0.400
Sulphuric acid,	0.180

Portions of this soil, on being dried in an oven at 300 deg. lost thirty-four per cent. of water. The silex is extrmely fine, and similar in appearance to the Hyde county soils. It is, however, in a greater proportion, and there is less organic matter. But there is no doubt this soil will be productive when drained and put under cultivation. It appears established from observation and experiment upon the swamp lands of the eastern counties, that much depends on the fineness of the earthy matter; for when there is a perceptible coarseness, the land will not bear cultivation many years. There is in those cases, however, less alumina and iron, and hence this kind of soil dries readily; and in certain seasons crops will be very short, and in reality fail. Where the earthy matter is fine it retains moisture, and furnishes a supply for those seasons when the rains are unseasonable. In certain cases the extreme fineness of the earth would present other defects. It would become too compact and close, and exclude the air. But the vegetable matter counteracts this defect in the swamp lands.

The gallberry lands often appear rich, if their vegetation did not remind one of their poverty. It will be found, in most cases of the poorest kinds of this class of lands, that the sand may be seen in the mass, or shows through its black covering of vegetable mould. On examination, the sand will be found to be coarse. Under cultivation the vegetable matter disappears rapidly; it is readily burnt—and the surface soon becomes white with the marine sand, and in extreme cases blows into ridges. Lands of this description do not pay the expense incurred in draining. It is sometimes necessary to drain them, in order to effect the drainage of other contiguous tracts.

Neither of the four foregoing soils of Gen. Blount's plantations belong to the poor gallberry lands, though No. 2 might be ranked in the better class of this description of soils.

The texture of the gallberry lands has much to do with their poverty; for generally they are made up of stiff whitish clays and coarse sand. From analysis we might prove that their constituents were the same as in productive kinds of soils. Such facts prove that productiveness is not entirely dependent on composition.

CHAPTER III.

Topography of the Eastern Counties, from Wake eastward to Onslow County. Character of the soil of the White Oak Desert. Mr. Francke's Pocosin and Swamp Lands. Better kind of Gallberry Swamp Land. Swamp Lands of the Brown Marsh. Green Swamp Lands. Mr. McNeil. Will pay for drainage. Barren soil of Bogue Sound, furnished by D. A. Humphrey, Esq., with his letter. Cause of barrenness in these soils.

§ 21. From Wake county eastward to the shore of the Atlantic the country slopes gently, the greatest inclination being of course on the western side of the plane. Between Wake and Johnston the country is rolling. From Smithfield, in Johnston, to Clinton, in Sampson county, the country is still somewhat rolling; but much less so than between Johnston and Wake. A large proportion of the country, however, between Smithfield and Clinton is a flat piney woods. The land seven or eight miles west of Clinton is level and rather sandy.

In Duplin county the level swamp lands begin. Between Magnolia on the railroad and Onslow county, the country is low and swampy, and in Onslow there are large tracts of unsettled or unreclaimed swamp and pocosin lands of an excellent quality. One tract in particular contains a hundred square miles, and a large proportion of it is excellent swamp land—and some tracts are equal to the corn lands of Hyde county.

Johnston county contains large tracts of flat piney woods, the soil of which produces only the shrubs which indicate unproductiveness, as the gallberry, ilex, and magnolia or bay, with a small growth of the long leaved pine. The surface, if not covered with water, is liable to be overflowed—and as it consists of sand and clay, with a mixture of vegetable mould, may be said to be quite impervious to water; and hence, the surface water stands over it for a long time, and its temperature remains too low for the growth of the more valuable trees and plants. Towards Sampson county the country improves, and upon the branches of the Six Run there are rich plantations. The best swamp lands are still farther east; and these, while they are usually high enough to admit of drainage, are rarely more than fifty feet above tide level. The Hyde county corn lands are about five feet above tide level, or may be less than four feet. Sometimes, in close proximity to the sounds, as in Carteret, the swamps are heaped up as it were, and hence may be from twelve to sixteen feet above the level of the sea.

In Onslow county, the soil between Thompson's and Jacksonville is very good. Some of it is suitable for the ground pea, being a light soil with considerable vegetable matter.

§ 22. In Onslow, the White Oak desert is the most interesting tract of swamp land in the county, it is at the head of White Oak creek. This tract may be drained into Trent river. The timber is very large, and consists of white oak, poplar and pines.

The most important work which has been undertaken, is the drainage of a part of this tract by Mr. Francke. He has been able to secure two objects, the drainage of the land and a good water power, with a fall of about twelve feet. The cost of cutting the main drain or canal is fifteen cents per square yard. The thickness of the soil in Mr. Franke's pocosin* is five feet towards the outer rim, and still thicker towards the middle, attaining at least ten feet of rich soil. This pocosin is said to vary much in its depth and quality; some parts are sandy, and the trees are still large and numerous. These sandy knowles are called *islands*. But the excellent quality of parts of it which are covered with heavy timber, prove by cultivation that it is equal to the Matamuskeet lands of Hyde—their average yield being twelve barrels of corn to the acre.

I have not seen the land referred to in Jones county, but I am confirmed in the statement from its composition, which I have determined by a careful analysis. Thus the drained portion of Mr. Francke's pocosin gave me a result on analysis equal in value to the best of the Hyde county soils. It is as follows:

Silex,	60.000
Organic matter,	25.000
Peroxide of iron and alumina,	11.030
Phosphoric acid,	0.312
Lime,	1 500
Magnesia,	0.300
Potash,	0.010
Soda,	0.020
Soluble silica,	0.100
Water,	2.713

From the foregoing results, when compared with those obtained by anaylsis of the Hyde county soil, it will be acknowledged that if composition is a test which can be relied upon, the Onslow swamp lands must be very valuable; and furthermore, that this value justifies the expense required in draining. This is the first question to be settled in all swamp lands: are their qualities good enough to justify this necessary expense? because they must be drained before the cereals can be cultivated. The encouragement to incur this first expense arises from the fact that when drained they do not

* This pocosin is partly in Onslow and partly in Jones county. The portion which has been drained and cleared is in Jones county. The only meaning which I can attach to the word pocosin is, that it is a large swamp.

wear out in the life time of man; they require no manures, they are easily tilled, and they produce large crops annually, and besides are less affected by droughts; or, in other words, the corn crop is more sure and certain than upon up lands.

Where there are large continuous tracts as in Onslow, Jones, Hyde and Beaufort, a systematic plan of drainage should be undertaken. This should be based upon a topographical survey of the whole tract, ascertaining first the area and its irregularities, if any, then the regular slope and the most feasible points to which the drains and canals should run. If a main canal can be cut which will take water sufficient for boat navigation, it should be regarded as an important means for transportation. It is surprising that swamp lands hold so much water—so that most of the largest tracts of pocosin lands furnish a sufficiency for this purpose.

The earthy matter in the pocosin of Onslow is very fine, and of drab color, in which respects it is similar to the best lands of Hyde.

It is evident also from an inspection of the results of this analysis, that there is a full supply of lime, and of the more expensive elements, and hence it may be expected that when these lands have been brought under full cultivation by thorough drainage and other means necessary to favor the growth of the cereals, that farms or plantations as valuable as any in North-Carolina, will be formed out of this desert swamp. The determination of the high value of this part of Onslow I consider of great importance; for there seems to have been hitherto great backwardness in attempting to reclaim the lands of White Oak desert. It is true the undertaking is a formidable one, but the rich results which will certainly be secured thereby fully warrant the undertaking.

§ 23. The character of the gallberry lands require also new investigation. These have usually been regarded as worthless. They are usually flat and wet, and hence the temperature of the surface is always too low for the vigorous growth of the most valuable trees: aside from this fact it is probable that the soil is really poor and unfertile, and no measures within a reasonable expense could be employed to change this semi-barren condition to one of fertility. But it is equally probable that many large tracts of land which are classed among the gallberry lands may be reclaimed and will become fertile by thorough drainage.

In forming a judgment upon the expediency of draining these flat and wet lands with a view to their cultivation, it is necessary to examine the texture of the materials which compose them as well as their composition. As there is a large proportion of black vegetable matter upon the surface, it is important to ascertain if it is intermixed with earth, and if so whether it is coarse or fine, and whether it is mostly sand, whose particles are large or visible at once on inspection. If the earth, after the vegetable matter has been consumed, is fine and impalpable, it is a fact which speaks well of its character; if on the contrary it is a white and coarsish sand, it is unfavorable, for it cannot be expected that it holds, in mechanical combination the more essential earths, alumina, lime and magnesia, or the alkalies, potash and soda. If it is sand these important elements will be in combination with the vegetable matter, and when this has become an ash, or is partly consumed, the soil will be destitute of the elements of fertility. Observation and experience prove the correctness of the foregoing observations. If, for instance, the soils of Hyde county are examined, the fine impalpable material is always found intermixed with the vegetable matter; and so, in cases where the sand is found, and soon appears after cultivation, the lands do not wear well but soon give out.

But the gallberry lands are frequently stiff, whitish clays intermixed with sand. These have undergone very little change from the influence of atmospheric agencies. When ploughed and exposed for a few years to the atmosphere the color slowly changes to a light brown, and finally to a deeper. These changes are also favorable, and it will be found that these lands improve by cultivation.

As an example of the better kind of gallberry land, I propose to give the composition of one which occupies a large area in Onslow county, which, on being submitted to analysis, gave the following results:

Silex,	82.300
Peroxide of iron and alumina,	8.700
Lime,	0.020
Magnesia,	0.010
Phosphoric acid,	0.150
Organic matter,	3.350
Potash and soda,	traces,
Soluble Silica,	0.100
Water,	6.000

The color of this soil is a light yellow, and its texture rather fine, and is disposed to be lumpy. Its texture and composition favor the growth of wheat rather than corn, and I have no doubt when reclaimed by drainage will prove an excellent soil for the cultivation of this grain.

§ 24. The swamp lands of Brunswick and New Hanover, and the adjoining counties, resemble in many respects those of Hyde and Onslow. In order to determine as far as possible from analysis the expediency of draining a certain tract or a portion of it lying in Brunswick county, which is known as the Green swamp, Mr. McNeil* furnished me with a few samples of muck which were obtained as it appeared from beneath the water. It was similar to black mud, but on drying I found it contained partially decayed pieces of bark, wood and roots, though its structure did not appear to be fibrous.

On drying in the paper in which it was orginally wrapped, it became rather hard and firm, showing that it contained earth, for if made up of peaty matter destitute of earth, it would have been much less firm and compact.

On submitting this material to analysis, I found it was composed of the following elements:

---

*Jack Forest, 24th November, 1857.

Dear Sir:—I send you four packages of soil from our swamp lands: one from the heavy timbered land on the Brunswick marsh; one from the low lands of the Brown marsh, and lands requiring ditching; one from the original Green swamps, but now timbered with young growth, and one from a ditch draining the land near the swamp, which I suppose contains lime.

Yours truly,

H. J. McNEIL.

Silex,	35.350
Peroxide of iron and alumina,	10.85
Organic matter,	37.50
Water,	15.8
Lime,	1.40
Magneisa,	0.15
Potash,	0.10
Soda,	0.15

This soil was found to be much richer than I anticipated, and on drying in paper, it retained a larger quantity of water than I expected. If the composition had been obtained after most of the water was expelled by heat, the proportion of the elements of fertility would have been proportionally greater. As the soil is composed, there can scarcely remain a doubt of the value of these lands. The earthy matter is as fine as that of the Onslow or Hyde county lands, and its quantity and condition proves, as it appears to me, the same capability with them for a productive cultivation for a series of years. Hence the cost of drainage should be incurred, and these valuable lands reclaimed, inasmuch as they pay better than the uplands. The extent of unreclaimed lands of this description makes it still more expedient, inasmuch as the general results are proportionately greater than when the surface embraces only a few acres.

The depth of this material is from eighteen to twenty-five or thirty inches, but like the Onslow pocosin it is variable, and like the latter also, the swamp abounds in islands, which are frequently occupied by inhabitants who contrive to live by basket making. The timber consists of cypress and black gum, and various pines and oaks, which frequently attain a large size, proving by the natural method a productive soil. In passing through these low lands, the water is frequently deep in the common highway; sometimes it is due to the prevalence of rains, in others it is produced by dams to obtain a water power for mills. As it respects the practice of maintaining mills in this low and half inundated country, it seems to me to be inexpedient. It certainly prevents in part the reclamation of these lands by drainage, and when it is taken into consideration that steam power cannot be very expensive in a country abounding in wood, it becomes quite plain that all such mills should be suffered to go down and their places supplied by the much more efficient steam mills.

The soil taken from the bank of a ditch is of a dark drab or purplish gray. It coheres strongly on drying and loses most of its water. It is gritty to the feel and is composed of moderately fine quartz and clay. On submitting it to analysis I found it composed of

Silex,	83.00
Organic matter,	21.20
Peroxide iron and alumina,	7.40
Lime,	trace,
Magnesia,	trace,
Potash and soda, undetermined,	
Water,	3.20

The lime and magnesia were scarcely perceptible. It resembles in appearance and composition the poorer gallberry lands, though it is probably better than many. If a soil of this description was to be put under cultivation it would require steady and constant marling. It forms a good subsoil in one respect, that of being impervious and capable of holding manures. It unlies the cultivable soil of the swamp lands in this neighborhood. The soil taken from the Brunswick swamp is brown or brownish; contains undecomposed twigs, bark, &c., but on drying forms a firm mass and contains a sufficiency of earthy matter. It is not unlike much of the soil of Hyde county, and it appears that it has been heavily timbered. I found it composed of

Silex,	45.470
Water,	8.000
Organic matter,	34.000
Peroxide of iron and alumina,	10.490
Lime,	0.490
Magnesia,	0.060
Potash,	0.581
Soda,	0.326
Soluble silica,	0.580

This soil possesses all the good qualities of the Hyde county soils. It absorbs and retains water strongly. The mass of soil on drying becomes hard and tough, requiring force to break it, and yet when apparently perfectly dry holds eight per cent of water. It is also sufficiently rich in lime, and particularly in organic matter. The question to be solved by analysis was whether these lands would become valuable by drainage. We may be assured this is proved by the results obtained by analysis. The expediency of drainage depends, however, very much upon the cost of the undertaking, but if the lands admit of drainage at the ordinary cost of such undertakings there is no doubt but that the soil would rank among the most valuable in the State.

§ 25. The foregoing analysis furnish examples of soils, most of which may be regarded as highly productive. In the midst however of productive lands, there are very frequently limited tracts which are really barren, so far as the cereals are concerned. To the eye, or upon a mere cursory examination, these tracts would be regarded as valuable as any which lie adjacent to them; yet experience would prove, in an attempt to cultivate them, that they are worthless. Corn takes root and grows a few weeks, when it begins to turn yellow, and finally dries up, or lives on in a stinted condition.

The cause of this unexpected termination is not well understood. Some planters believe that the soil is lacking in one or more of the elements of growth; others, that there is some substance of a poisonous quality in the soil. If either of these suppositions or guesses were true, the fact might be determined by submitting the soil to a careful analysis.

But there are other causes which affect unfavorably the growth of vegetables. It may be too tenacious, it may be compact and prevent the access of air, (an element always required,) or it may be so porous and open that the necessary amount of moisture cannot be retained. In addition, therefore, to the chemical composition of a soil which a plant may require to insure its perfection, there may be an incompatible physical one, whose operation is equally effective in stinting its growth. We must not, therefore, regard barrenness as always the result of the absence of fertilizing elements. In investigating any particular case of infertility, it is necessary in the first place to inquire into its physical condition— to ascertain its texture, the size of its particles, and at the same time ascertain whether they are silicious and coarse, and insusceptible of retaining water or fertilizing matter.

Many examples of these unproductive tracts belong, geologically, to the most recent formation, as the Postpliocene of authors. They are properly marine formations, in which sand, as will be seen in the sequel, forms the largest proportion of the elements of the compound.

A specimen of the unproductive soil was received from D. A. Humphrey, Esq., of Swansboro', Onslow county, accompanied with a letter containing a brief account of the material under consideration, the copy of which is in the following words:

SWANSBORO', N. C., *Jan.*, 1858.

DEAR SIR:—You will remember, that at Beaufort, last May, when I had the pleasure of an introduction to you, you told me if I would send you a specimen of some of that peculiar land of which we talked, you would analyze and inform me of its constituents, and advise me of the necessary change to be made in it, so as to make it produce the ordinary crops.

The land from which this specimen was taken produces weeds and vegetables common to all the sound land, very scantily, except the sweet fennel (Fœniculum) which grows very luxuriantly, so large even, that I have them taken up with a grub-hoe. It will produce, with the best cultivation, (without manure,) say 100 lbs. seed cotton to the acre, and one bushel corn. When the corn first springs up, it grows rapidly for a short time; then turns yellow and falls. The land is quite elevated. I have shipped to Wilmington a small bag containing the specimen, from which place you will soon receive it, and when it suits your convenience to examine, please do so, and let me hear from you.

And oblige, very much,
Your friend and humble serv't,
D. A. HUMPHREY.

PROF. E. EMMONS, Raleigh, N. C.

On submitting the soil described in the foregoing letter, I found it composed of the following elements:

Silex,	85.200
Peroxide of iron and alumina,	2.862

Carbonate of lime,	1.85
Magneisa,	trace,
Organic matter,	7.05
Water,	2.20

The phosphates and potash scarcely distinguishable in 200 grains. The sand representing the silex is rather coarse, grains distinctly visible and rather angular. The color of the mass is black, and it seems to be made up of fine vegetable matter. It contains, as will be seen, a sufficient quantity of lime and inorganic matter—the former is derived from particles of marine shells, sometimes of a large size, and it is probable all the lime is coarse; it effervesces with acids. The silex, though large, is not in greater proportion than in many productive soils. It would be regarded as a light soil, though the vegetable matter might deceive one who has had no experience in cultivating soils of this description. A soil of this character presents two questions for solution: 1st, whether its present or natural state will justify an expenditure sufficient to make it fertile? and 2d, if so, what course should be adopted to secure the object sought for? My first impression is that it cannot be made productive at all, in consequence of its composition. It has really only a base of coarsish sand of considerable depth. Hence it is loose and porous, and transmits all the water through it. Besides, it is evident that there is a deficiency of alumina and all the most expensive elements except lime, and the lime, instead of being fine and in a condition to furnish to vegetables this necessary element, aids rather in giving it porosity, as it is in coarse particles. But still, so far as this element is concerned, the soil is well enough; but in a combination or mixture which is loose and porous, it is doubtful whether the necessary chemical changes do take place at all. considering the nature of the tract of land, I believe the first step to be taken towards its improvement would be to give it a heavy dressing of clay, to change, if possible, its physical condition. Less clay would be required, if one which is calcareous could be employed; for less would answer the purpose than if it were pure. In order that chemical changes should take place, it is necessary that water should be retained, or that it should pass through slowly.

The fertilizers which are best adapted to a case like the Swansboro' soil are green crops, peas or clover, which may be ploughed in. By either crop we secure in part the end we aim at, condensation of the soil or compactness, by which water is retained, and by which also time is given for the consummation of the chemical changes required. The water being retained, the crop, whatever it may be, the plant is supplied both with water and nutriment.

But the necessary dressing of clay is always expensive, even when it is near or at hand, unless indeed it can be reached by the plough. There are very few cases where the expense of hauling clay is ever returned in an increased amount of crops. We may be able, as I believe, to point out in what way given defects in a soil may be remedied.

When that is done, it still remains a question for solution, whether the mode proposed will pay. It is evident that a calculation of the cost of the mode prescribed is very important, if it is to be put in execution. A garden may be put into a high state of fertility, when a large cornfield cannot be treated in the same mode.

It is not easy, in the case before us, to account for the barrenness of the soil of the coast, unless we adopt the theory that it is mainly owing to its mechanical condition. A soil having a very close resemblance to this, at Cape Cod, in Massachusetts, is quite fertile. President Hitchcock, of Amherst College, who conducted the geological survey of the State, found on examination and analysis, that the blowing sands of the cape owed their productiveness probably to the comminuted shells, intermixed with the sand. Or, at least, the sands, under a microscope, exhibited particles of shells; and hence, as the soil consisted of sand and finely comminuted shells, its productiveness was attributed to the presence of this fine lime dust commingled with the sand. But the climate of Massachusetts bay is much more moist and cool during the summer than the coast of Bogue sound. The sun in the latter case acts with more force upon vegetables than at the north. A soil which might bear corn in Massachusetts would not sustain it on the coast of North-Carolina, on account of the more rapid evaporation of water; in consequence of which, a plant would be early deprived both of water and nutriment, though it might be found in the medium in which it had been growing.

CHAPTER IV.

Soils of Jones county, taken from the plantation of J. H. Haughton, Esq. Composition of a brown earth overlying and resting upon the marl beds. Recapitulation.

§ 27. Several specimens of soil have been furnished me for analysis from Jones county, which, as they may be employed to illustrate the composition of the cultivated lands in that section of the State, I shall give the results in this place. They were furnished by John H. Haughton, Esq., from a plantation which he recently purchased. Four kinds were forwarded, marked 1, 2, 3, 4 respectively. No. 1. Color, brown or blackish brown, and to the eye appears rich in vegetable matter. When ignited it loses readily this part of the soil and becomes a light drab, leaving a fine residue resembling that of the Hyde county soils. Its appearance shows that it is a silicious soil. One hundred parts gave me the following proportions:

Silex,	82.300
Peroxide of iron and alumina,	4.300
Organic matter,	4.500
Lime,	0.102
Magnesia,	0.020
Potash,	0.003
Soda,	0.001
Water,	8.800
Sulphuric acid,	trace,
Chlorine,	trace,
	100.026

This soil has evidently been worn by long cultivation, still it has sufficient matter to sustain moderate crops; but it has reached that stage which requires additional applications of manure.

All the most important elements, as phosphoric acid, sulphuric acid, lime, magnesia and potash, are considerably less than the standard soils contain; and as they maintain about the usual proportions to each other, it is probable that they have been reduced simultaneously by cultivation.

No. 2. Color, a light drab, resembles clay, but contains coarse particles of sand, and hence is very gritty. This variety of soil contains greater excess of sand, and is deficient in organic matter, etc. One hundred grains gave me

Silex,	93.000
Peroxide of iron and alumina,	2.000
Organic matter,	1.300
Lime,	0.001
Magnesia,	0.010
Water,	3.000
Potash,	trace,
Soda,	trace,
Sulphuric acid,	trace,
	99.311

This evidently ranks among the poorest of soils. It appears quite similar to much of the poor gallberry lands of the eastern part of the State.

A larger proportion of alumina and iron could have been obtained by fusion with baryta or soda, but the exhaustion by boiling with hydrochloric acid, I deemed sufficient for my purpose, or the objects to be obtained by analysis. This kind of soil no doubt might be put into a condition for raising wheat by thorough drainage, and a large application of manures.

The best application to a soil, the composition of which resembles the foregoing, is a compost of marl with organic matters derived from the stable; or, the leaves of a forest. In materials of this description a supply of organic matters is obtained in combination with the phosphates of lime and

5

potash, all of which are required to impart fertility to a soil defective as this is in each of those elements.

No. 3. Color, brown, fine grained, and has apparently considerable vegetable matter in its composition. It has no lumps of earth, but is reduced to a granular state; or in other words it is pulverulent and light.

One hundred grains, on being submitted to analysis, gave me—

Silex,	89.300
Alumina and peroxide of iron,	2.550
Lime,	0.151
Magnesia,	0.020
Phosphcric acid,	trace,
Sulphuric acid,	0.020
Potash,	0.001
Soda,	0.C02
Organic matter,	3.100
Water,	3.000
	98.144

The quantity of organic matter is less than its appearance before analysis indicated, and this is often the case in the soils in the eastern part of the State.

Many chemists regard the organic matter as of little importance. Experience and the best conducted experiments, however, prove that it is a necessary constituent of a good soil.

Here, also, the lime or alkaline earths and alkalies are deficient, at least to raise good crops of maize, or any of the cereals. Besides there is a great excess of silex, but it is in a fine condition, indeed in none of the samples is it ever coarse; it, therefore, makes a better basis upon which to work than if this were a coarse sand, inasmuch as it is better conditioned to hold or retain water.

No. 4. Color, nearly black, with organic matter, and fine grained. Ignition leaves it of a drab color.

I found its composition, on submitting it to analysis, to be as follows:

Silex,	88.700
Peroxide of iron and alumina,	3.350
Lime,	0.100
Magnesia,	0.022
Sulphuric acid,	0.010
Chlorine,	trace,
Potash,	0.048
Soda,	0.010
Organic matter,	1.800
Water,	5.000
	99.040

This specimen of soil has a better composition than either of the four of this lot. There is less silica, more lime and potash; though the amount of organic matter and peroxide of iron and alumina is still comparatively small, and we infer from that fact, that the amount of phosphates is also small.

This soil has no doubt been under cultivation for years. It has a good basis to build upon, as the silex is fine and not very excessive in quantity. It is evidently a better soil than No. 1, and does not rank in the class with No. 2, which is a coarse clayey silicious soil, the particles of which are very coarse. In all these samples the cultivation should not be carried to that extent which would effect an entire exhaustion.

The remarks upon the four foregoing soils have been suggested by the analyses and their physical properties. No information has been obtained respecting the treatment to which they have been subjected.

§ 28. A soil of a somewhat remarkable appearance, and having a good composition, is spread over large portions of the eastern counties. It is not always a surface soil; indeed it is rather rare to meet with it under cultivation. It occupies a distinct position in the series of soils, and is really one of the deposits which is always associated with the marl beds. It cannot, with propriety, be regarded as a marl, though under favorable circumstances it may be used as a fertilizer.

It has a brown color, and when wet is as tenacious as the ordinary clays, though it has less alumina in its composition; it is very adhesive to the shoe or boot, and if it is ever profitable to haul clay for fertilizing the sandy soils, this is especially adapted to the fulfilment of all the ends which may be obtained by the use of clay.

It rests upon the shell marl in some places, and in others upon the eocene marl. The circumstances attending its deposition were peculiar. It appears to have been deposited immediately after a period of denudation, as it rests not only upon the marl, but extends into, and fills deep channels which had been cut out of the marl during the period alluded to. Hence it appears to send down long tapering columns which extend sometimes to a point near the bottom of the bed. This formation, however, was formed from quiet waters, as there is no evidence of a rush or violent flow of waters, by the presence of large rocks, or even coarse pebbles. It has some coarse sand intermixed with pebbles. It has the appearance of a sediment, which was probably derived from the decomposing slates and granite, which lie beneath the tertiary, but which is now concealed, except in a few isolated places.

On submitting this soil to analysis I found it composed of

Silex,	77.850
Alumina and peroxide of iron,	10.107
Lime,	2.000
Magneisa,	1.810
Organic matter,	3.950
Water,	5.750
Sulphuric acid,	0.010
Chlorine,	0.010
Potash,	0.185
Soda,	0.345
Soluble silica,	0.100
	99,815

This soil is rich in lime, which is in part derived from a few small fragments of shell which it contains, but it effervesces but slightly, and hence it is probable the lime is diffused rather uniformly through the mass. When this mass lies immediately beneath the sandy soil, and within reach of the plough, it would improve it very much to commingle it with the surface material, and it need not be rejected in loading marl at the pit, inasmuch as its composition shows that it is an important improver of the common sandy soil so prevalent in the eastern counties.

The phosphoric acid remains to be determined. In itself this soil has a composition admirably adapted to the growth of wheat, or indeed cotton. It contains also a large amount of potash.

It was taken from a mass which overlies the eocene marl of the plantation of Sam'l Biddle, Esq., of Craven county. It is, however, found on the Cape Fear, resting upon the shell marl, a more recent deposit, and may be found on the plantation of Dr. Robinson, of Elizabethtown.

RECAPITULATION OF THE LEADING FACTS RESPECTING THE SOILS OF THE EASTERN COUNTIES OF NORTH-CAROLINA.

§ 29. (1.) The soils of the eastern counties, without exception, are marine formations, being deposited from water, and are truly sediments. They are therefore in their origin unlike those of the middle and western counties, inasmuch as the latter are the products of slow decomposition, and are *in situ*, or occupy the place upon the rocks from which they are derived.

The eastern soils have, on the contrary, been transported, or were first the products of a disintegration and, afterwards, transported from the places from whence they were derived. As they are frequently composed of one or, at most, two materials which can be distinguished by the naked eye, it is impossible to determine the source from whence they came. They were probably derived, however, from the granite which borders the tertiary formation upon the west. Their distinguishing features are siliceous; and it seems that most of the aluminous compounds, as felspar and certain slates, were finely comminuted, and were transported to distant points, leaving the heavy and coarser materials in the bays which jut up from the ocean in the depressions of the land.

These sandy deposits were not laid down at one period, though they are comparatively modern. They alternate with a few beds of clay, but there is but one near the surface which is extensively distributed. The last of the marine deposits was mostly a pure white sand; and it not unfrequently washes white when it is deprived of its vegetable coating. The last or most recent bed of sand, is formed by waves of the ocean into swells or undulations. A belt thus thrown up and moulded by this agency, extends obliquely across the country. One of the most distinguished features of this belt is intersected by the Wilmington railway, at Everettsville, ten miles S. W. from Goldsborough. These swells of sand are sufficiently large and extensive to give origin to permanent mill-streams. They seem to have been derived from the Atlantic side, and to have been cast up by waves which in their operation have denuded all the eastern portions lying between this belt and the Atlantic ocean, and hence it not unfrequently happens that the upper stratum of sediment is a stiff clay.

(2.) The denuded clay is often a stiff brick clay, and is about four feet thick. Shallow depressions are hollowed out of it, which are always the receptacles of water, and have also favored the growth of moss and small vegetables. To the growth of these humble plants we attribute the origin of the vegetable matter which is so extensively prevalent in many of the eastern counties, and which are known by the names of *pocosin* and *swamp lands*.

(3.) A slight elevatory movement of the whole coast of North-Carolina, has reclaimed those tracts from water; and, though not dry yet, they are not submerged, and are no longer the recipients of sediment.

While these lands were but half reclaimed from the dominion of water, they were subjected to inundations which transported fine silt, and which required much time to settle. This fine silt, or mud, is now the soil which is so productive in corn in Hyde county and other parts of the Atlantic border.

This singular soil is characterized by its vegetable matter, and by the extreme fineness of its inorganic matter; and the two compound elements are well fitted to each other, and admirably adapted to the growth of maize in this climate, whereas in a northern climate it is very doubtful whether the same results could be obtained. In Canada East there are somewhat similar soils, but they are treated quite differently in order to bring the soil under cultivation. There, the surface is first burned, and the ash and debris remaining supplies the nutriment for a succession of heavy crops. When this first fertilizing matter, obtained by burning, is exhausted, it is subjected to the same treatment again, and again put under cultivation. The lands of the eastern counties would not bear this mode of cultivation; neither do they require it. They become productive by draining.

§ 30. The composition of the soil of Canada East, taken from a tract which is there known by the name of *Savanne* of St. Dominique, is composed, according to Mr. Hunt, of

Fixed carbon,	29.57
Ashes,	6.75
Volatile matter,	63.68

The ash or inorganic matter in 100 parts contained:

Carb. Lime,	52.410
Lime, ... } as silicates,	10.430
Magnesia, }	3.150
Peroxide of iron,	4.680
Alumina,	2.440
Oxide of magnesia,	0.040
Phosphate of lime,	2.019
Sulphate of lime,	15.085
Sulphate of potash,	0.605
Sulphate of soda,	0.076
Chloride of Sodium,	0.412
Silica,	4.920
Sand,	4.040
	100.308

In the foregoing analysis we can readily perceive that the material subjected to this process is an ash, with only faint traces of soil, but in appearance the North-Carolina pocosin lands resemble the turf or peat soils of Canada and New York, but the better kinds or those of Hyde, contain, intermixed with the vegetable matter, fine earth, which gives them a substantial body. In this respect they differ from the peaty or turf soils of other places. They differ also in endurance. They continue productive through several generations. Those of Hyde have been tilled through three generations, and the fourth has them under culture. I attribute this extended period of endurance to the temperature which the soil enjoys. Below, in immediate proximity to the roots of corn, the water remains through the season. Hence there is a temperature preserved which is only moderately high in the midst of summer, in consequence of evaporation. Even the water often surrounds the hill of corn, and remains on the surface for a long time, without injuring the growth of the plant. The external heat is sufficient for the crop. If it were higher it would slowly consume the vegetable matter. Besides, the low temperature of these peculiar soils, the proximity to the ocean, favors a constantly moist climate, or atmosphere; and hence, through the influence of water beneath, and a moist atmosphere above, the growth of vegetables is promoted.

In the midland counties the vegetable matter is consumed, or so nearly consumed that the blackened belt at the surface is never formed. Upon the mountains, the whole of the blue ridge, vegetable matter accumulates in the soil. The heat is insufficient to destroy it, while in the midland counties it never accumulates even in forests, and though there is a large annual addition of vegetable matter from the leaves which fall in autumn and winter, still no accumulation takes place in the soil. It is literally consumed.

§ 31. The pocosin and swamp lands present a great variety in the proportions of vegetable matter present in the soil. Some passing to the extreme limit, from 10 to 93 per cent. of organic substance. The latter percentage is near the boundary which limits the capability of growing the cereals. A greater excess of vegetable matter scarcely admits of the continued growth until the crop ripens, it soon ceases to grow, becomes yellow after it has appeared above the ground when it has reached the height of 10 or 12 inches. The most valuable swamp and pocosin lands lie in Hyde, Beaufort, Jones, Onslow and Brunswick counties; those of Hyde have been steadily cultivated for more than one hundred years without manures, and still the crops are equally as good as when first planted. Hundreds of square miles of the most valuable of these lands still remain unsubdued. It may be inferred that, as these swamp lands are so low and wet, that they must necessarily be extremely unhealthy, or become so when drained and the vegetable matter begins to decompose. Experience, however, does not support this view. The testimony of those who have cultivated them for forty years is, that their families have enjoyed as much health as their neighbors who have lived at a distance. Persons who are in the habit of plunging into the swamp lands knee deep for draining, and when drained to live in the immediate vicinity of the extended surface of black vegetable mould for years, are rarely sick with fevers. The points which are unhealthy are those which are exposed to winds which blow over extended surfaces of the waters of the Neuse or Cape Fear rivers. Miasm, which generates fever, arises more from the banks of rivers than from the swamp and pocosin soils.

§ 32. The soil which is known as the gallberry soil is not of a uniform composition or appearance; one of the most common kinds is formed of sand, intermixed with black vegetable matter. On exposure to rains by the road-side, or where ditches are cut through it so as to expose a section one or two feet thick, it has a grayish look from the presence of the white marine sand which is exposed by washing. A microscope shows at once the naked sand. A soil of this description, and which is widely spread over the flat low grounds of the middle section of the eastern counties, I submitted to careful analysis for the purpose of determining the amount of available material which it contains. It was taken from the plantation of Mr. Lane, of Craven county, but is a fair representation of the soil of the Dover pocosin. It contained:

Sand or silex,	70.50
Organic matter,	25.20
Peroxide of iron and alumina,	0.76
Lime,	0.01
Magnesia,	trace,
Water,	2.70
Soluble silica,	trace.

The silex is a perfectly white marine sand.

Although this analysis is not carried through, yet it is evident that the available matter for crops is extremely small. The seventy-six hundredths of a grain of peroxide of iron and alumina is too small a quantity to have much chemical or mechanical influence upon the organic matter with which it is mixed; neither can it furnish phosphoric acid to supply the wants of vegetation if put under cultivation. This variety of gallberry land belongs to the poorest class of soils. It is not expected it would pay a profit if cleared, and hence all such lands should remain wild, or in their natural state.

Another variety of low ground soil is of a better quality, though still it ranks low for the purposes of agriculture. It is of a light color, and hence contains much less vegetable matter. It is a marine sand, intermixed with a small quantity of clay, a portion of which can be dissolved in hydrochloric acid. This soil is from Sampson county. It forms extensive areas in Johnston, Sampson and Duplin counties. There is, however, an improvement in the character of the low grounds towards the east from Johnston county. The color of this soil is a light brownish or purplish drab; in drying it becomes hard and loses most of its water of absorption. It resembles the green swamp soil in Brunswick county. It is composed of

Silex,	88.40
Peroxide of iron and alumina,	2.92
Lime,	0.02
Magnesia,	0.03
Water,	3.09
Organic matter,	4.20
Potash and soda,	traces,
Phosphoric acid, undetermined,	

In this variety of soil from the swampy grounds there is still a deficiency of the alkalies and alkaline earths; this, however, may be cultivated with medium results, if marl is at hand from which to supply the deficient matter.

CHAPTER V.

FERTILIZERS.

What constitutes a Fertilizer.—Sources of Fertilizers.—Those from the Vegetable kingdom are the Ash.—Ash of differnt Vegetables —Ash of Plants resembles in composition the Inorganic Matter of Soils.—Quantity of Fertilizing Matter removed from the Soil by different Plants.—Methods to be adopted by which a Waste of Fertilizing Matter may be Prevented.—Fertilizing Matter Restored by Plowing in Green Crops.

§ 33. Any substance in husbandry is a fertilizer which improves the soil. They are numerous and are derived from numerous sources. The air is a reservoir of substances which improve the soil, and water is the medium of communication. As in the laboratory substances do not act upon each other unless one or both are in a fluid condition; so fertilizers must be in solution in a menstrum, of which water, in the kingdom of nature, is the universal solvent ' The air contains ammonia and carbonic acid. These are the most direct fertilizers. They are both transferable agents, passing from the atmosphere to the earth dissolved in rain water, and escaping upward from the earth in the ascending vapors, when they have fulfilled their mission to the grown and perfect vegetable. They escape when it decays, and wait for another mission to the earth or soil. The interchange is almost perpetual. There are vegetables at all times undergoing decay, or [*eremacausis*,] a slow combustion, during which the compound atoms are undergoing a change, and each one of which is finally resolved into new forms and conditions. Ammonia and carbonic acid are the common products of change in all these cases. Both are, however, compound bodies. The first is a body recognized by its extremely pungent smell, and commonly known as hartshorn, and is formed by the union of two elements—nitrogen and hydrogen. The latter is the lightest substance known—it is .069, the weight of air. Carbonic acid is an air, also, or gas, and is heavier than atmospheric air, and hence is sometimes found in depressed places, not, as is usually maintained, by falling down from the atmosphere in consequence of its greater weight, but by its escape from beneath, or from the soil or fissures of rocks. Rain water and snow hold both ammonia and carbonic acid in solution, and hence, as has been remarked, they are the media from which growing plants derive these important fertilizers. Snow, particularly, is rich in ammonia. From this material it may be obtained by evaporation. To this substance, probably, the beautiful greenness of vegetation is due, which appears on the melting of a March snow.

These two substances, however, may be derived from any organic matter in the earth, when it is undergoing decay; hence, most if not all bodies which have lived may furnish them if buried in the soil and within reach of the roots of a growing plant. There are, therefore, two modes by which these fertilizers become subservient to nutrition—1, by water falling from the atmosphere and, 2, by water in the soil which dissolves them out from particles of earth and organic matter.

In the application of the first mode, husbandry has nothing to do. It is a part of the machinery of nature, by which she maintains the balance between the vegetable, animal and mineral kingdoms. This machinery in its workings is perfectly competent to preserve this balance, to furnish food and sustain in perpetual existence all the species which belong to the present system. In a temperate climate, however, without artificial aid, the cereals would cease to grow, or yield the harvests they now do, because of the exhaustion they bring about in the progress of time and of cultivation.

§ 34. Fertilizers may be divided into kinds according to

the source from whence they are derived, as those which belong to the three kingdoms of nature, the mineral, vegetable and animal, but such a division is really of small importance, inasmuch as it will be perceived from the foregoing remarks that all fertilizers may be traced back to the mineral kingdom, even ammonia is strictly a mineral, although it abounds in both the vegetable and animal kingdoms in certain combinations. Proximately, they are either animal or vegetable; but in either case they are of a mineral origin. The fertilizers which will come up for examination are ashes, marls, excrements of animals and green crops.

§ 35. It needs no argument to prove the value of ashes as fertilizers, we have only to inspect the foregoing tables of the composition of the ashes of wheat, maize, oats and potatoes. The composition of the ashes of forest trees brings us to the same results, and as much dependence is placed upon the decomposition of the standing trees in the cultivated fields it is important that the fertilizers thus obtained may be shown. We are obliged, in this case, to resort to the analyses of the ash obtained directly by combustion. The results, however, are the same in the natural process of decay as by combustion, and the decayed bark, limbs and twigs furnish ultimately what they would have furnished were they consumed by fire.

The white oak, for example, *quercus alba*, furnishes by cumbustion an ash composed of the following elements. First the bark of the trunk, which contains:

Potash,	0.25
Soda,	2.57
Sodium,	0.08
Chlorine,	0.12
Sulphuric acid,	0.03
Phosphates of lime and magnesia,	10.10
Carbonic acid,	29.00
Lime,	54.89
Magnesia,	0.20
Silica,	0.25
Soluble silica,	0.25
Organic matter,	1.16

The bark of the twigs gave me, on submitting the ash to analysis:

Potash,	1.27
Soda,	4.13
Chlorine,	0.13
Sulphuric acid,	trace,
Phosphates of lime, magnesia and peroxide of iron,	14.15
Carbonic acid,	30.33
Lime,	47.72
Magnesia,	0.20
Silica,	0.65
Soluble silica,	0.55
Organic matter,	1.52
	100.09

The wood of the twigs decays with the bark, but the wood, as will be seen, is richer in fertilizing matter than the bark. It has the following elements:

Potash,	9.74
Soda,	6.89
Sodium,	0.16
Chlorine,	0.25
Phosphates of lime, magnesia and peroxide of iron,	23.60
Carbonic acid,	17.45
Lime,	34.10
Magnesia,	0.50
Silica,	0.55
Soluble silica,	0.60
Organic matter,	5.90
	99.99

The outside wood slowly decays beneath the bark, or after it has fallen and furnishes an ash rich in potash and the phosphates of lime, magnesia, etc. While standing the process is certainly very slow, but it will ultimately be reduced to a substance equivalent to an ash having the following composition, viz:

Potash,	13.41
Soda,	0.62
Sodium,	2.78

Chlorine,	4.24
Sulphuric acid,	0.12
Phosphates of lime, magnesia and iron,	32.25
Carbonic acid,	8.95
Lime,	30.85
Magnesia,	0.36
Silica,	0.21
Soluble silica,	0.80
Organic matter,	5.70
	100.18

The pine tree gives an ash on combustion differing slightly from the foregoing, viz:

	BARK.
Potash,	2.86
Soda,	3.17
Chloride sodium,	0.03
Sulphuric acid,	3.48
Carbonic acid,	24.33
Lime,	31.48
Magnesia,	0.01
Phosphate of lime, magnesia and peroxide of iron,	22.12
Organic matter,	3.58
Silica,	13.40

The most important addition which the bark of this species of pine will add to the soil is soluble silica and lime, the alkalies are comparatively unimportant.

§ 36. The benefit which has been attributed to the standing dead trees is not probably due entirely to the ash which the bark and limbs furnish. A more important effect may be obtained by the moisture which is retained by the spreading roots in the soil, each of which must absorb considerable water and retain it for a long time. The practice adopted in this particular is better adapted to a warm than a colder climate. The shade even of the trunks of forest trees would be detrimental to the maize crop in New England or New York, more, as I believe, than all the benefits to be expected either from its decaying wood or the increased water in the soil.

The leaves of forest trees are richer in the phosphates than the bark or wood.

In the fruit these elements exist in still greater proportion. In the leaves of the Catawba grape I found them to exist in the following proportions:

Potash,	13.394
Soda,	9.698
Phosphates of lime and magnesia,	32.950
Lime,	4.391
Magnesia,	1.740
Chlorine,	0.740
Sulphuric acid,	2.620
Silica,	29.650
Carbonic acid,	3.050
	99.026

The fruit of the common black butternut is composed of

	RIND.	SHELL.
Potash,	41.43	47.00
Soda,	7.12	10.21
Earthy phosphates,	15.60	18.50
Lime,	23.75	5.60
Magnesia,	1.55	0.10
Chlorine,	1.50	2.15
Silica,	1.35	0.40
Sulphuric acid,	2.65	9.84
Org'ic matter and alkaline phosphates,	2.30	5.40

§ 37. The oat plant furnishes similar facts. The dry crop in the grain weighs 975 lbs. per acre, and furnishes 39 lbs. of ash, with a percentage of 4.00. The elements, per acre, are:

Phosphoric acid,	6.00
Sulphuric acid,	0.40
Chlorine,	0.20
Lime,	12.00
Magnesia,	3.00
Potash and soda,	5.00
Silica,	21.00
Oxide of iron,	0.60

In the straw, per acre, the proportion of elements is:

Phosphoric acid,	1.50
Sulphuric acid,	2.50
Chlorine,	3.00
Lime,	5.00
Magnesia,	15.00
Potash and soda,	17.00
Silica,	24.00
Oxide of iron,	1.00

§ 38. The clover-plant weighs, when dry, 3693 lbs. per acre. The percentage of ash is 7.70, which is quite large, and the weight of the ash, per acre, 284 lbs. It contains, of

Phosphoric acid,	18.00
Sulphuric acid,	7.00
Chlorine,	7.00
Lime,	70.00
Magneisa,	18.00
Potash and soda,	77.00
Silica,	15.00
Oxide of iron,	0.90

The clover plant, it will be perceived, contains about equal proportions of lime, potash and soda; the lime, however, is in excess, but its composition shows why it is so well adapted as a fertilizer to the wheat crop. The vigorous growth of clover upon a soil which has been marled with green sand, which contains both lime and potash, illustrates and places in a strong light the advantages of special fertilizers.

If the ash of the foregoing, or any other plant is compared with the composition of tho best soils, or marls, it will not fail to strike almost any one that there is a close resemblance between them. The soil furnishes phosphoric acid, iron, sulphuric acid, chlorine, magnesia, silica, potash and soda. All the remarkable fertilizers contain the same elements. Those which are the most striking in their effects contain lime, phosphoric acid, potash and soda in large proportions, furnishing thereby the expensive elements, the most essential ones, or those which exist in the soil in the smallest proportions, in great abundance. The effects of a fertilizer are the most perceptible where these are the most abundant. Hence guano which contains a large amount of phosphoric acid, ammonia and lime, rarely fails to satisfy the wants of the plant and to become the efficient means of producing a greatly increased crop. Of certain elements it may be said there is never a deficiency. Silica is one, as it is always present in the largest proportion. The same may be said of iron; but lime, magnesia, and especially the alkalies, are frequently wanting, if not altogether, yet not in a sufficient quantity to supply the wants of vegetation. Hence, in fertilizers, the test of their value consists in determining the quantity of lime, potash and phosphoric acid, which they contain; or, the amount of those special elements which are always in the smallest proportion in the soil; and hence too it is easy to perceive why soils become barren by cultivation, as those elements are early removed in the crops which the soil has borne.

§ 39. To illustrate this point and make it sufficiently clear to be comprehended by every reader, I propose to state the quantity of nutriment which several of our most important plants consume; and which is derived directly from the soil.

In order to do this it is necessarry to ascertain what elements exist in the plant, and which must of necessity be taken from the soil in which it grows. These elements are obtained when a plant is burned. The residue of the combustion are earths, intermixed with alkalies, the mass of which is known as ashes; wheat, oats, potatoe and clover, will furnish striking examples suitable for the illustration of the point in question.

An ordinary wheat crop, according to Bousingault, when dried, weighs, upon an average, in grain, 1052 lbs.; in straw 2558 lbs., and the grain furnishes 2.40 per cent of ash, and the straw 7.00. The quantity of ash per acre, in the grain amounts to 25 lbs., in the straw per acre is 179 lbs.

The proportion of the elements contained in the 25 lbs. of ash are:

	LBS.
Phosphoric acid,	12.00
Sulphuric acid,	0.30
Chlorine,	trace,
Lime,	0.80
Magnesia,	4.00
Potash and soda,	7.00
Silica,	0.04
Oxide of iron,	0.00

In the straw the proportions are:

Phosphoric acid,	5.00
Sulphuric acid,	1.50
Chlorine,	1.00
Lime,	15.00
Magnesia,	9.00
Potash and soda,	17.00
Silica,	121.00
Oxide of iron,	1.75

One remark may be made in this place, that the phosphoric acid of the grain greatly exceeds that of the straw; while the lime of the straw is in much greater proportion than it is in the grain, and the silica is reduced in the grain to the smallest percentage, but greatly abounds in the straw. We have in this, as in many other instances, the exercise of a species of elective affinity, by which the elements select their appropriate organic materials.

A potatoe crop, when dried, weighs, in tubers, 2828 lbs., and gives, in ashes, 4 per cent., and weighs 113 lbs. per acre. The percentage of composition is:

	LBS.
Phosphoric acid,	13.00
Sulphuric acid,	8.00
Chlorine,	3.00
Lime,	2.00
Magnesia,	6.00
Potash and soda,	58.00
Silica,	6.00
Oxide of iron,	17.00

The percentage in tops, 5042 lbs., with 6 per cent. of ash,

and weighing 303 lbs. per acre. The percentage of composition is:

	LBS.
Phosphoric acid,	33.00
Sulphuric acid,	7.00
Chlorine,	4.00
Lime,	7.00
Magnesia,	5.00
Potash and soda,	135.00
Silica,	39.00
Oxide of iron,	16.01

The potatoe plant abounds in the oxide of iron and potash, and there is no doubt the character of the soil influences to a considerable extent the quality of the tuber.

§ 40. Among the substances which of all others would be expected to be destitute of inorganic matter are cotton wool, and the fine fibre of flax. Indeed it was at one time maintained that these substances were composed of carbon, oxygen and hydrogen, and hence would be entirely volatilized by heat; and hence, too, as they were composed of those bodies, their cultivation would not impoverish the soil, provided the other parts were duly returned to it. But these views proved fallacious. Prof. Shepard, on submitting the cotton wool to analysis several years ago, found the percentage of the ash to be 0.9247, nearly one per cent. The ash, as obtained, gave the following results in his analysis, viz:

Carbonate of potash, (traces of soda,)	44.19
Phos. of lime, (traces magnesia,)	25.44
Carbonate of lime,	8.87
Carbonate of magnesia,	6.85
Sulp. potash,	2.70
Alumina, (accidental,)	1.40
Chlorides, potassium and magnesium, Sulp. of lime, Phos. potash, oxide of iron and loss,	6.43

This analysis is quoted for the purpose of showing that the finest fibre contain matter derived from the soil. So of the finest flax fibre whose ash is found to contain:

Carbonate of lime,	62.00
Sulphate of lime,	7.15
Phosphate of lime,	13.66
Oxide of iron,	3.99
Carb. of magnesia, with traces of chloride of sodium,	2.00
Silica,	11.20
	100.00

The steep water in which flax is rotted contains a small amount of matters dissolved out of the flax, but neither the addition to the soil of this water, nor the refuse of its dressing is sufficient to restore the soil to the state it was in prior to the growth of the crop.

§ 41. Various methods are adopted to supply the waste in fertilizing matter, or to diminish it during cultivation. One of the cheapest methods is to allow as much of the crop to decay upon the field as possible.

This course is adopted when a planter ploughs in the stalks of indian corn, cotton, or the stubble of rye and wheat. There is an advantage in ploughing in the stubble of all cereals. Another method has been adopted. The stubble is first burned and the ashes have been strewed over the field under the impression that they contain all the fertilizing matter. This method, however, has never proved successful. This is due in part to the nature of the ash. All silicious stems, when heated to redness and burned, undergo, so far as their silica is concerned, an important change, which consists in converting the soluble into an insoluble silica, and is therefore not immediately available to the plant; when ploughed in entire and allowed to waste in the soil, all the soluble silica is preserved in a condition to meet the wants of the growing vegetable.

The plants which belong to the corn family, however, are not so profitably employed as fertilizers as clover, buckwheat and the pea. This fact becomes obvious from an inspection of the composition of the corn stalk, or the stubble, or straw of wheat, and comparing it with the composition of the latter. Still, the use of the corn stalk is highly important. I have found it composed of the following elements:

Potash,	16.210
Soda,	24.699
Phosphates of lime and magnesia,	15.150
Lime,	2.820
Magnesia,	0.936
Silica,	12.850
Sulphuric acid,	10.793
Chlorine,	10.453
Carbonic acid,	1.850
Organic matter,	3.200
	99.461

§ 42. The inspection of the composition of the ash of the corn stalk shows that it should not be wasted, inasmuch as a quantity of the most valuable elements would be lost; it would be equivalent to the wasting of so much bread or corn, inasmuch as the whole of the matter may be converted into bread or corn in the process of cultivation.

The straw of wheat is less rich in phosphates and the alkalies than corn; and yet it is entitled to preservation and use as a fertilizer.

The ash of the straw amounts to 2.660 per cent., and consists of

Silica,	1.235
Phosphates,	0.422

Thus the phosphates bear a very small proportion to the silica.

The complete analysis of the straw of wheat gave me:

Potash,	22.245
Soda,	5.195
Earthy phosphates,	19.600
Silica,	49.100
Lime,	3.460
Magnesia,	0.324
Sulphuric acid,	0.876
Chlorine,	0.121

In a ton of straw the loss which would be sustained by wasting it, amounts, in pounds, to

Silica,	29.255
Potash,	13.253
Soda,	3.095
Earthy phosphates,	11.678
Lime,	2.061
Magnesia,	0.193
Sulp. acid,	0.521
Chlorine,	0.072
	60,128

The organic matter, which is not taken into the account, is equally valuable and important, both as furnishing materials of growth and the preservation of an open condition of the soils.

§ 43. Certain crops are raised expressly for the purpose of improving the soil. These, when in blossom, are ploughed in, and their subsequent decay furnishes the manure for the succeeding crop. The kinds usually selected are those which grow vigorously and send their roots deep. Such plants bring from a great depth the fertilizing matter to the surface where it becomes accessible to the succeeding crop.

The red clover is the favorite plant in the Northern States. Buckwheat is also employed, but it is objectionable: it continues to spring up from the seed as some will ripen and mix with the wheat crop or appear as a weed in the corn and require eradication by the hoe.

For the South the pea has become a favorite with intelligent planters, and is, from its composition and adaptation to climate the best crop to precede wheat and to act as its fertilizer.

The composition of red clover is well adapted to the end which it is designed to fulfil; besides, its root is large, spreads widely and sinks deeply, and hence it brings to the surface a large amount of fertilizing matter.

The ash of the green plant amounts to 1.06 per cent., when dry to 5.87.

On submitting the dry clover in the condition of hay to analysis I found:

Potash,	25.930
Soda,	14.915

Earthy phosphates,	20.600
Carb. of Lime,	30.950
Chlorine,	1.845
Sulphuric acid,	0.495

If a ton of this hay or a plant in its green state was ploughed in, it would add the following amount of elements reckoned in pounds as follows:

Potash,	32.153
Soda,	18.394
Earthy phosphates,	25.544
Carbonate of lime,	38.378
Magnesia,	4.873
Chlorine,	2.288
Sulphuric acid,	0.624
Silica,	1.054
Amounting to	123.308 lbs.

§ 44. It is not perhaps possible to estimate the real value of a clover crop as a fertilizer. Two hundred pounds of guano cost $5. May we not infer that its value exceeds that of this popular fertilizer, especially when it is considered that the organic part must exercise considerable influence and always furnishes a large amount of food? It is true that new elements are added by the clover, but then the cost of the crop is trifling, and the effects are more lasting than guano in this climate.

The clover crop is from two and a half to three tons per acre of dry hay. It is more profitable to feed cattle upon it before it is ploughed. By this course or plan of treatment the manure which is added by feeding cattle nearly suffices for the diminished amount of clover consumed. It is not regarded as expedient to plough in a very heavy green crop of any kind. It is better to feed it in part, if there were no valuable returns in meat or flesh.

On account of the grain in food for cattle the clover crop is preferable to buckwheat, and yet this plant is rich in fertilizing products.

§ 45. In the South the heavy or large stalks of corn are

broken down and laid flat and longitudinally with the furrow and covered in that position.

The cotton stalk is also laid flat and ploughed under. The real importance of this operation becomes evident on an inspection of the composition even of the dried stalks, bolls or capsules.

I found from the composition of the capsules that they are richer than the stalks.

The percentage of ash of the dry capsules is 5.402, nearly six per cent. It was obtained from capsules left in the field growing in the county of Nash.

§ 46. The ploughing in of the dry plant returns a certain amount to the soil. From the capsules there will be returned in every hundred parts of ash of percentage of ash 5.60:

Earthy and alkaline phosphates and potash,	21.480
Soda,	5.230
Earthy phosphates,	22.923
Lime,	31.940
Magnesia,	11.627
Sulphuric acid,	0.400
Chlorine,	0.231
Soluble silica,	1.302
Adherent sand,	2.601
	97.734

In the stalks of cotton in the condition in which they are broken down preparatory to ploughing the field I found the following elements:

Alkaline and earthy phosphates,	14.400
Potash,	17.400
Soda,	20.860
Lime,	31.200
Magneisa,	13.160
Sulphuric acid,	3.046
Chlorine,	0.400
Soluble silica,	0.100
	100.566

§ 47. From the foregoing analysis it is evident that the

custom of ploughing in the old stalk after the cotton is saved is an important measure.

I have no means of determining the number of tons of the stalks per acre, but the amount thus saved to the soil or succeeding crops is very great and prolongs the fertility of a cotton plantation for years.

In this connection it is proper to state the composition of the cotton seed, which is now always employed as a fertilizer. Its real value will be duly appreciated, though it is scarcely necessary to confirm by analysis what experience had long determined by its use. But the planter will understand better what he is adding to his soil, and also how much from the following results of analysis;

Earthy phosphates,	32.000
Potash,	15.560
Soda,	10.960
Lime,	4.000
Magnesia,	0.200
Sulphuric acid,	2.720
Chlorine,	0.120
Carbonic acid,	8.540
Soluble silica,	2.000
Adherent sand,	23.600
	99.700

The large quantity of sand is due to cotton adhering to the seed which had been exposed in a pile to the weather. It was not suspected until the ash was subjected to the action of hydrochloric acid. It is of course foreign matter.

After making all the allowance necessary for this foreign matter it will not fail to strike every cotton grower of the value of the cotton seed as a fertilizer.

§ 48. Analysis of the seed of buckwheat:

Potash,	21.27
Soda,	2.32
Phosphoric acid,	49.85
Lime,	3.01
Magnesia,	15.84
Sulphuric acid,	1.55

Silica,	1.95
Chlorine,	0.30
Carbonic acid,	1.95
Organic matter,	2.75

In the cultivation of this plant it will be seen that a large amount of fertilizing matter is removed in the gathering of seed, or, if it remains, a large amount is preserved for subsequent crops.

Every ten bushels of seed contains 6.281 lbs. of phosphoric acid, two pounds of magnesia, and over two pounds and a half of potash. The whole amount of valuable fertilizers removed in every ten bushels of buckwheat is 12.450 lbs. The buckwheat in drying loses about the same quantity of water as wheat and rye. Thus, on being dried in a water bath at 212, it lost 12.875 parts; and hence there remains of dry matter, 87.125 of which gives 4.132 per cent. of ash.

The organic constitution of buckwheat is similar to the cereals, consisting of

Starch,	42.47
Sugar and extractive matter,	6.16
Dextrine,	1.60
Epidermis or insoluble matter,	16.42
A peculiar gray matter, soluble in potash, but insoluble in water or alcohol,	10.10
Albumen,	6.70
Casein,	0.78
Oil,	0.47
Water,	12.88

§ 49. The foregoing does not relate so much to matters which can be employed as fertilizers, but is introduced here for the purpose of showing its nutrient properties.

The pea will no doubt take the place of the red clover in this State. Experience has already proved its superiority. It is easily cultivated and is not liable to so many accidents. It takes deep root and spreads widely, and is rich in valuable fertilizers. By careful extraction from the hill I have found its roots spreading through six feet of ground.

That the value of the pea may be appreciated, and its fertilizing matter applied to the best advantage, I have carefully determined the composition of its ash from specimens which I obtained in Wake county.

The percentage of ash of the pea vine, destitute of leaves and in the condition in which it is fed to cattle, and as derived from 268 grains of the stems and branches in a perfectly dry state, I found to be 4.570.

On submitting this ash to analysis I found it composed of

Potash,	7.800
Soda,	5.650
Earthy phosphates,	19.800
Lime,	16.400
Magnesia,	30.040
Sulphuric acid,	11.710
Chlorine,	1.710
Silica,	10.900
Soluble silica,	6.000

When we find so large a percentage of ash, and a composition clearly rich in inorganic constituents, we may not doubt the utility of employing this plant as a fertilizer instead of the clover plant, as it is considerably richer in the expensive elements of nutrition.

§ 50. The pea with its pod is richer in phosphates than the vine, and as these are ripe when turned under the value of the crop for this purpose is increased.

The percentage of ash, as determined from 365 grains of the dried pod with the pea, is 3.13. The percentage of ash is greater from the presence of the pod. But this being ploughed in the result is more accurate from their combination. If the nutrient matters of the pea were to be determined it should be analyzed by itself.

The composition of the pea, with its pod, I found as follows:

Potash,	24.200
Soda,	10.789
Earthy and alkaline phosphates,	32.200
Carbonate lime,	11.000
Magnesia,	3.000

Sulphuric acid,	1.461
Chlorine,	0.561
Soluble silica,	10.020
Silica,	3.800
Percentage of ash,	3.137

The pea in composition is closely related to the cereals, and in nutritive powers ranks high. Indeed the leguminous plants as a class stand at the head of a certain class of nutrients. The bean employed for food gives more muscle or strength of muscle and endurance than the cereals. This is due in part no doubt to its phosphoric acid and nitrogenous matters.

It appears from the foregoing that the greater the amount of nutrient power the more valuable they are as fertilizers. Weeds which bear only small seeds, or which are composed of lime, are less useful than leguminous plants, and others which are closely related to the cereals.

§ 51. The composition of another plant which may be interesting in another point of view is tobacco. I design to show by the analysis how much the tobacco exhausts the soil, and of what elements.

Thus, one hundred parts of the ash consist of

Potash,	4.260
Soda,	6.140
Lime,	48.000
Magnesia,	9.180
Phosphates of lime and magnesia, etc.,	14.300
Sulphuric acid,	8.420
Chlorine,	1.100
Silica and sand,	4.800
Soluble silica,	3.800
	100.000

This tobacco grew in Rockingham county, and was regarded by the manufacturer as fine as any which is grown in the northern counties. The result, however, of this analysis surprised me, as it contained so much less of potash than can be expected in the best of tobacco. It is found by many analyses, however, that the ash is variable in the proportions of its elements.

The tobacco which obtains the highest price in the Paris market contains a much larger proportion of potash and less lime. This specimen had the fine yellow brown color which is regarded as indicative of the best quality. As it is, however, it is a lime plant, nearly one-half being composed of carbonate of lime.

CHAPTER VI.

FERTILIZERS—CONTINUED.

Marl beds, or Marl formations.—The different periods to which they belong, or their relation to each other.

§ 52. There are three distinct formations from which marl is obtained. Enumerating them in the ascending order, or according to age, they lie relatively to each other as follows: 1. *Green Sand;* 2. *Eocene Marl;* 3. *Miocene Marl.*

The first, or green sand, is the formation which is so favorably known in New Jersey as a fertilizer, having been employed for that purpose for more than half a century. It derived its name partly from its green color, and partly from its granular consistence. The beds thus named are known not only in this country but also in many parts of Europe by the same name, and where, to a certain extent, they are also used as a fertilizer.

In the geological systems its beds are subordinate to the cretaceous system, and in Europe form subordinate beds beneath the chalk—the white chalk in common use for marking.

In this country this part of the cretaceous system is wanting, or has not yet been recognized. From its wide extent, both in this country and Europe, it is, geologically speaking, an important formation; so also in an economical point of view it is equally important, for it has been a source of revenue to the agricultural community, not second even to guano. For permanent improvements in the soil it is superior to this far famed substance, its effects lasting from ten to fifteen years. In New Jersey it first attracted attention from an accident: some green sand being thrown out of a ditch upon a bank, an exceeding fine growth of clover was the consequence. It was immediately inferred that the substance upon the ditch bank was the cause of this fine growth; and hence a trial was made of it.

From many subsequent experiments and observations its claim as a good fertilizer became established. This happened more than fifty years ago, and ample experience in the mean time has fully satisfied the agricultural community at large that it is worthy the confidence which has been reposed in it.

§ 53. In the subsequent pages I propose to give a full statement of the grounds upon which its reputation rests, and also to furnish numerous analyses of the best and poorest varieties of this substance. In the first place I deem it proper to show its geological relations, and its relative position to other beds of marl, inasmuch as it will aid in determinining in any given case whether the substance or beds in question really belong to those which have received the common name referred to. In all cases this is an economical question, or may be thus used, inasmuch as the beds formed during this geological era have a composition which fits them for the purpose for which they have been so largely employed. Beds, therefore, occupying their position may be supposed without trial and without analysis to contain the active fertilizing matter. It, however, cannot be determined by these external observations, how much they contain, for it is found that they are variable in composition, so far as quantity is concerned. For the purpose of determining their commercial value, or to ascertain the amount which may be profitably employed and how far they may be transported has to be ascertained by analysis.

There are several localities at which the green sand occurs. The strongest marl beds occur at Black Rock on the Cape Fear river, about twenty-five miles above Wilmington. It forms low bluffs at several other points, but it appears to terminate from two to five miles below Brown's landing.

Striking across the county to the eastward it again appears prominently at Rocky Point, twenty miles above Wilmington. The green sand, unlike the shell marl, forms continuous beds, but as its beds are undulating, they rise at certain points to the surface, and then sink beneath it.

In this State I have been unable to determine its thickness, or the number of beds which properly belong to it. For this reason I propose to describe them now, as they are known to exist in New Jersey, inasmuch as such a description may aid others where it exists, to determine with accuracy both their thickness and the number of beds which compose the green sand formation in North-Carolina. The difficulty in the way of solving this question is the slight elevation of the banks of rivers and ravines above the adjacent country. We find at Black Rock, for example, a strong bluff of this deposit, but the water is never low enough to disclose the bottom beds, or the masses upon which it rests.

In order to state all that is known of the green sand and marl, and their relations to each other, I have prepared several sections which show how they are situated with respect to each other. From these sections it will be seen that the marl beds vary much in thickness, and in their relations at different places where they are exposed to the best advantage. Thus, section I, fig. 1, exhibits all the beds as they exist at Black Rock:

Fig. 1.

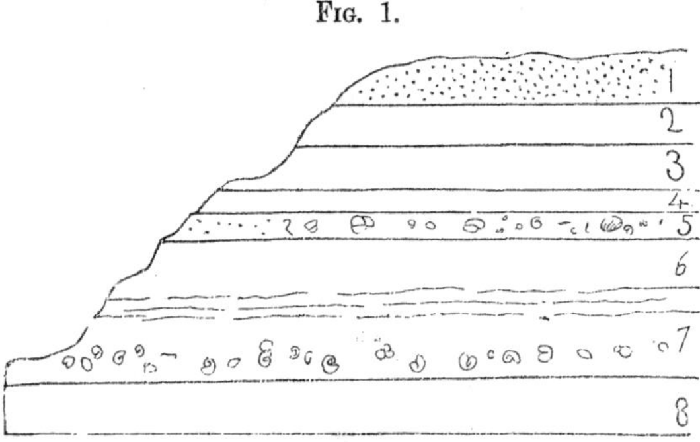

1. The upper bed is the common marine sand spread widely over the county. 2. Beneath it there is a mass of brown soil, or earth, which is probably more widely spread than any other in the eastern part of the State. It is sometimes pebbly towards the upper part, and at many places the pebbles are cemented by oxide of iron. A pudding stone is thereby formed, which is very firm, and has been employed as a rough building material. In the vicinity of Fayetteville it is not unfrequently used for the more ordinary kinds of construction. From the vicinity of Raleigh eastward it may be seen by the road-side where a cut has been extended through the superincumbent sand. This bed, which is at least twelve feet thick at Fayetteville, originated in the decomposition of primary rocks, the debris of which becomes red, or reddish brown, by exposure to the atmosphere. If any thing, it is more persistent towards the belt where these rocks formed the surface materials. How this stratum has been spread out so evenly and widely through the whole width of the State from south to north is not satisfactorily accounted for. Along the western margin referred to it rests on the rocks from which it is derived. Eastward, however, where recent beds of different kinds take their proper places, this brown earth formation is found near the surface, but with several marine strata beneath and upon which it reposes. It always maintains the position I have given it, or its relations are never altered; and hence, though it may be regarded as a soil, still it must

have been spread out by some general cause, and at one specific period.

This bed, however, is not confined to this State. It extends over a part of Maryland, Virginia, South Carolina, Georgia, and Alabama.

It is, therefore, a wide spread stratum, having its origin through the influence of general causes. That this cause or force operated with considerable violence is indicated by the losses which one at least of the inferior formations has sustained. The shell marl, for example, is never a continuous deposit, and some of the beds are frequently furrowed and channelled, apparently by a rush of water over them, removing not only the upper layers, but cutting frequently deep into the beds. An erosion of this kind is illustrated by fig. 5. The brown earth fills these eroded channels without mixing at all with the marl.

The next stratum beneath is a brick clay, which is also general, but it is absent occasionally, in which case the brown bed occupies its place. This clay varies considerably in composition; it is sometimes charged with sand, in others it is very fine and compact, and makes the best of brick. It passes also into potter's clay. It is bluish white, gray and reddish at different places. It never exceeds five feet in thickness.

4. The fourth stratum is sand, usually gray, and loose in texture, not unlike quick sand.

5. The shell marl occupies the fifth place in the descending order. It will be fully described hereafter.

6. The beds of green sand occupy the sixth place, and at Blackrock it may be divided into two beds; the upper contains a large amount of clay, and the lower is sandy with more lime; it is also indurated, or partially consolidated.

The lower mass forms a shelving projection from the upper, some eight or ten feet wide, when it falls off perpendicularly to a depth of fifteen feet. The lower part is always under water, and I know of no locality at which this part of the formation is exposed. I regard this as an unfortunate circumstance, inasmuch as I have reason to believe that the quality of the marl is better towards the bottom, or lower in the bank, than where it is exposed. At certain points in New Jersey it has a sandy base, but several feet above it becomes a rich marl.

The color of this kind of marl is green or dark green. It is always rather sandy, but still it is rich even then in fertilizing matter. The Blackrock beds here have a dark green, or greenish gray, and may be divided into two parts: the *upper* which has a darker color, and is much like clay to the feel; and the *lower*, which is consolidated and of a greenish gray, and rather gritty to the touch. There is no dividing line which is so clearly marked that we can fix upon the termination of the lower, and the beginning of the upper division, but still the difference observable is sufficiently strong to admit of the division I have proposed; though, geologically, it may be regarded as one mass. The division is more important in an economical point of view, inasmuch as the composition of the upper is quite dissimilar to the lower bed.

§ 54. In New Jersey the green sand formation is composed of six distinct beds; three of which are known as green sand proper, in consequence of the peculiar composition; and three which are composed of a common marine sand, and which separates each of the respective beds from the other. In North-Carolina it is probable that equivalent beds exist, but it has been impossible up to this time to recognize but two. At Blackrock the lowest is known by its fossils: the Exogyra costata, Ostrea falcata, Belemintes Americana, and casts of the cucullea vulgaris. This mass terminates in one which is quite argilaceous, and in this part of it no fossils have been observed.

The third or upper bed may be probably recognized at Tawboro', on the Tar river, at the marl beds of Col. Clark. It is only about four feet thick, but is underlaid by sand, in which much sulphuret of iron is disseminated.

The annexed section, fig. 2, shows the relations of the beds referred to upon the Tar river:

FIG. 2.

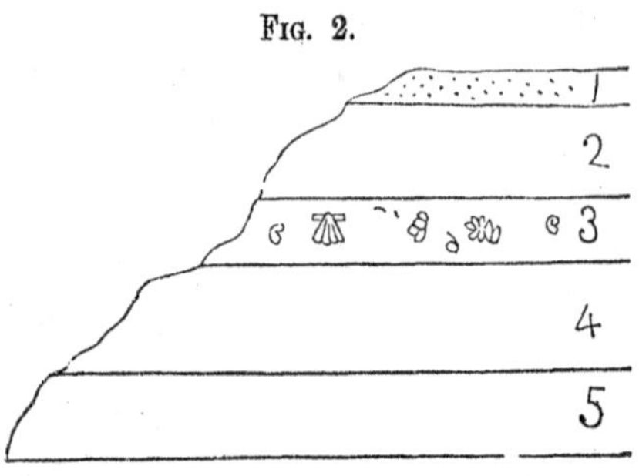

Soil. 1. Ten feet of yellow sand. 2. Four feet of greenish clay. 3. Six feet of shell marl. 4. Four feet of upper shell marl, containing lignite and pyrites. 5. Light gray sand, the thickness of which is undetermined, as it extends below the water of the Tar river, and does not become visible at any other place in the vicinity. It is probably one of the sand beds which seperate two of the adjacent beds of green sand. But as it has not furnished fossils it cannot be confidently maintained. It is, however, mineralogically, a green sand.

As all the beds of green sand are never exhibited at one place, and as those which have been spoken of, except the upper, on the Tar river, the thickness of this formation remains undetermined.

Wherever it occurs the country is comparatively low, and at no point yet discovered has the base of the Blackrock mass or lowest been sufficiently elevated to disclose, even approximately, its thickness.

§ 55. The bluffs which exhibit the tertiary and secondary formations of the eastern counties are mostly upon the south-side of the rivers and ravines. Some of these bluffs are high and commanding, but they are never continuous for long distances. The green sand does not appear in any bluff above Brown's landing. Indeed it disappears about three miles below, and though this landing is high and bold, yet I am unable to recognize a bed which can be referred to the upper part of the secondary formation.

At Brown's landing there are numerous distinct beds. In arrangement they belong to two distinct dates: 1st, the upper which is Miocene, and the lower which is probably Eocene.

These beds are exhibited in the following section:

FIG. 3.

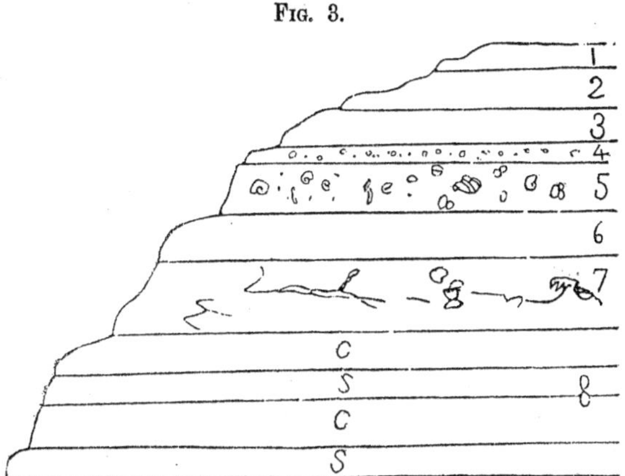

1. Sand. 2. Brown earth. 3. Clay, four or five feet thick. 4. Sand and pebbles. 5. Shell marl. 6. Sand, with consolidated beds which becomes a gray sandstone, with fossils and lignite. 7. Blue clay. 8. Sand, blue clay, succeeded again by sand. The formation below is here concealed under water.

The most interesting points at Brown's landing are the thick beds of sand and clay beneath the shell marl, the latter of which is identical with that at Black Rock, where, it will be recollected, this marl rests upon the upper bed of green sand. At the landing we find interposed at least sixty feet of material which does not occur at Black Rock at all. These intervening beds I regard as Eocene. It may, however, prove to be Miocene, and as a part of the lignite formation equivalent to that which is spread over large tracts of country in Nebraska and Kansas. It has consolidated beds, cemented by carbonate of lime, in which lignite is very common. Another fact of interest is the presence of green sand in the shell marl, while it is almost entirely absent in the inferior beds. The marl contains, also, Exogyra, Belemnites and coprolites which belong to the green sand which were washed from these beds. The change in passing from the Eocene to the Miocene was attended with considerable violence, as the latter have abundance of pebbles, rolled coprolites as hard as quartz, teeth, etc. The bottom is truly a pebbly bed.

§ 56. The sand beds beneath the shell marl extend nearly to Fayetteville. They may be examined at the bridge over Rockfish creek, seven miles from Fayetteville, and at Mrs. Purdy's marl bed, ten miles above Elizabethtown, and, also, at Elizabethtown, in the high banks below the village.

The sand of this formation, when it is unconsolidated, is loose and caves from its banks continually. In addition to lignite and a few shells it contains an abundance of iron pyrites. Its whole thickness on the Cape Fear is about seventy feet.

It is possible the beds may be recognized on the Neuse and Tar rivers, especially at the Sarpony hills, fourteen miles below Goldsboro'.

§ 57. The bluff below Elizabethtown presents the following strata, as exhibited in fig. 4:

FIG. 4.

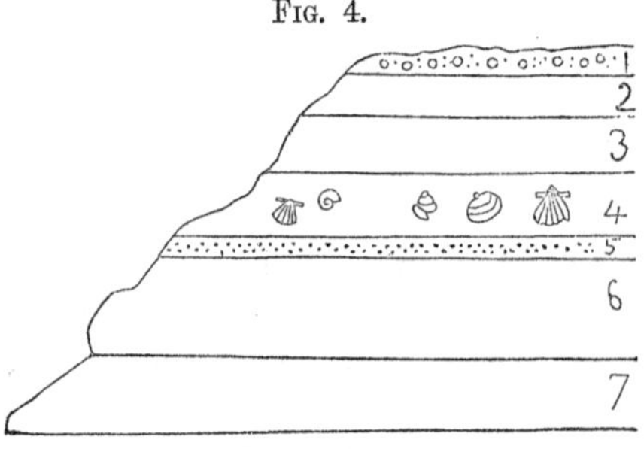

1. Sand with pebbles. 2. Brown earths. 3. Sand. 4. Shell marl three feet thick. 5. Sand containing lignite and consolidated layers, with numerous fossils.

The beds of sand with lignite or charred wood are similar to those of Brown's landing and Walker's bluff. But there are no particles of green sand or fossils from this formation in the shell marl bed. It appears that the shell marl beds in which are intermingled the organic remains from the secondary, are confined to a narrow belt which may be traced along the eastern border of the formation.

Section No. 5 is designed to show the relations of the shell marl to the white Eocene beds of the Neuse, which do not extend south-westward to the Cape Fear.

Fig. 5.

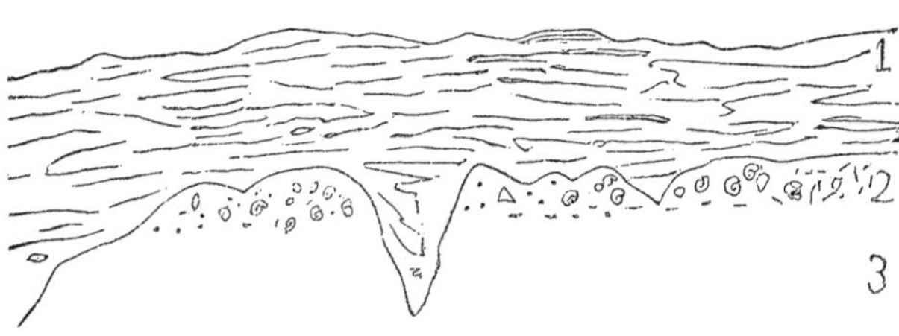

1. Soil, consisting of red earth penetrating into an excavation in the bed of Eocene marl. 2. Position of the ordinary shell marl. 3. Upper part of the bed in which most of the fossils occur. 4. Body of white, or light drab colored marl.

The section shows the marl beds of Mr. Wadsworth, of Craven county.

It will be observed that the shell marl is in contact with the drab colored marl, the entire mass of the lignite formation of the Cape Fear being absent. At this place, the brown earth is present filling the ancient fissures of denudation. The shell marl is not present at this point, but appears in the same relative position three or four hundred yards west from this bed.

§ 58. The foregoing sections show the diverse nature of the beds composing many of the bluffs of the Cape Fear, Neuse and Tar rivers. The same facts would be also shown by sections at many points upon the Roanoke and Meherrin rivers farther north. The position of the shell marl seems to change, as in one case it rests upon the green sand, in the second upon a lignite formation some sixty or seventy feet thick, and then again upon a whitish marl which is well known to belong to the Eocene period.

The formation above the shell marl is mostly a marine sand. Its thickness is variable, and it is sufficiently great to prove that a long interval had elapsed before the present was fully ushered in.

§ 59. The series of beds, from the green sand upwards, which hold a definite place in the geological scale, have been exhibited in the sections alluded to, do not take in the most recent. Upon the coast or near it I have observed limited patches of peaty deposits resting upon a marine sand, and upon the former beds of shells composed mainly, if not entirely, of those which now live upon the coast. These beds of shells are rarely more than ten or fifteen feet above high tide. The peaty beds, however, lie at the water's edge, and at many points are rapidly disappearing by the action of tides and waves.

The mode in which the shells are collected appears to have been similar to that which was instrumental in the accumulation of the common shell marl; they appear to be heaps of dead shells thrown up by the waves,—still they are perfect, or are but slightly worn by attrition. Those which are changed the most have become simply chalky from the action of the weather upon them since they were deposited. The beds which are now forming have received the name of *Eolian* by Lieut. Nelson. The sands of the entire coast come under this denomination, and may be regarded as deposits overlying the accumulation of beds of shells already alluded to.

§ 60. The formations then upon the coast and interior of N. Carolina may be subdivided into: 1. *Green Sand*, an important part of the secondary; 2. *Eocene*, consisting of white marl which is made up of comminuted corals and shells, and the lignite beds which consist of gray sand and pebbles, embracing consolidated beds and a few beds of clay; 3. *Miocene* or *Shell Marl*, which is composed of fragments and entire shells accumulated in banks; 4. *Pliocene* and *Postpliocene*, which are made up of peaty beds, banks of shells, and finally, moveable sands, (Eolian sands,) which are constantly moving beyond the present coast line. It should be observed, however, that the third or Miocene division is regarded by Prof. Holmes and the late Prof. Tuomey as Pliocene.

In this State I have obtained the same fossils in equal numbers as those in Virginia, where the beds still retain the designation, *Miocene*. Not only, however, do they contain the Virginia fossils, but those which in South-Carolina have served to change the name from Meiocene to Pliocene. It appears that many of the Virginia species belong to a warm climate, that they became extinct at an earlier period than at points farther south, and that the same species which were once common on the coast of Virginia and Maryland, and which are now extinct so far as that part of our coast is concerned, still live farther south where the climate is congenial to the species.

CHAPTER VII.

FERTILIZERS—CONTINUED.

Stone Marl, its economical value.—Composition of the Green Sand of the Cape Fear River.

§ 61. The marls of North-Carolina do not rank so high as the strong marls of other States. This is in consequence of the large proportion of sand with which they are intermixed. It appears that the coast has been from time immemorial the great depository of sand. The rivers from the interior carry sand or matter in which silex greatly predominates. The rocks in the interior belong to the silicious class. Limestones are absent. But the great amount of sand of the coast has been probably derived from more distant sources, and hence it is probable we must look to the regular currents of the ocean which flow in, more or less, upon it, for the determation of the source from which its sands have been derived. When the Atlantic tide reached inland as far as the last of the series of falls of the rivers of the State, as the Roanoke, Cape Fear and Neuse, it acted upon a granite rock which readily decomposed, and which must have furnished an immense quantity of silicious debris. This rock may, therefore, have been one of the sources of the sand alluded to. Some beds of marl are consolidated into rock, and where this consolidation was effected through the agency of soluble silica, it has become a durable mass, and fit for being used in building. It has received the name of *stone marl*, which I propose to speak of in the first place.

§ 62. *Stone Marl.* There are two varieties of stone marl, both of which deserve a special notice. The first consists of shells cemented strongly together, and which are usually from one to one and a half inches across, and very uniform as to size. They are very firmly cemented by silica, which seems to have penetrated the shells more or less. This rock has been employed for a long period for small mill stones. Its valuable qualities consist in being easily wrought when first removed from the quarry, but subsequently becomes very hard and strong. Being made up of shells, it has a rough appearance, even when cut evenly; but this feature constitutes its recommendation. For certain structures it is admirably adapted. The enclosure of the cemetery in Newbern is made of this rock, and the noble arches have an imposing effect. The rock is very durable, as appears to be well sustained by the rock itself, where it is exposed, or has been exposed for ages. For rough work it may be used without dressing, but for ornamental, if dressed properly, it is far superior to granite for all structures, where the material should be indestructible. It is adapted to the construction of dwellings, as the walls will continue dry in wet weather.

This rock underlies Newbern and the adjacent county. It extends fifteen or twenty miles in a northeast and southwest direction. In some places it reaches the surface; in others it is forty to fifty feet below. I regard it as one of the best building materials in the State.

The second variety is a granular cream colored rock, and rather destitute of shells. It might be mistaken for an oolite. The grain is uniform, and like the preceding is soft, when first taken from the quarry, but becomes hard as any rock after an exposure to the air for a few months. This rock is not disposed to disintegrate, and hence in this respect is superior to granite.

This granular variety occurs in Wayne county. The rocks

or consolidated parts of it are abundant on the plantation of Maj. Collier.

At a few places it is sufficiently pure to be burnt for lime; as a general rule it contains too much silex to make a strong lime.

The rock on Maj. Collier's plantation contains:

Silica,	59.400
Peroxide of iron in combination with alumina and phosphoric acid,	4.120
Carbonate of lime and a trace of magnesia,	36.480
	100.000

The amount of carbonate of lime is variable, and ranges in the consolidated varieties from 30 to 75 per cent. The silex in the rock exists in grains as sand, which are visible, but a soluble silica is no doubt the cementing material, which of course once existed in solution, or in a state of minute subdivision. This marl may be used in building, or if sufficiently pure and free from sand and silica, it may be burnt for lime, which will be adapted to agricultural purposes. Its composition fits it for this purpose as it contains a small proportion of phosphoric acid.

§ 63. The green sand is frequently partially consolidated, but never forms a building material. For agriculture, when the amount of potash is considered, it is the most important of the marls. In North-Carolina I have found no locality where its potash equals that of New Jersey. This I attribute in part to our inability to reach strata which are upon the same geological level, though it is probable that the amount of sand will be greater, and hence diminish proportionally the amount of available fertilizing matter.

The lowest mass accessible at Blackrock I found by analysis, has the following composition:

Silex and sand,	37.000
Peroxide of iron and alumina,	6.400
Carbonate of lime,	33.400
Phosphates of peroxide of iron,	1.600
Soluble silica,	1.460

Magnesia,	13.600
Potash,	1.431
Soda,	2.123
Organic matter,	1.600
Water,	1.800
	100.614

The sand is frequently in quite large angular grains. That part of the bed which is green, or properly green sand, is not so distinct as in New Jersey, and it would be impossible to separate the grains mechanically, while in New Jersey they may be separated from the other materials. These grains have been analyzed by Prof. Cook, who has found them composed of

Silica,	45.510
Protoxide of iron,	21.134
Alumina,	7.960
Magnesia,	2.400
Potash,	6.748
Lime,	3.842
Phosphoric acid,	0.990
Sulphuric acid,	1.129
Carbonic acid,	0.563
Sand,	0.850
Water,	9.110
	100.209

It has been found that the green grains in the green sand possess a very uniform composition, and that taking the average analysis of several specimens the grains contain silica, protoxide of iron, alumina, magnesia, potash and water in nearly equal proportions, while the other constituents are variable. The absence of the green grains in the marl of black rock may account for the small percentage of potash which is the principal element relied upon in the New Jersey marl. The lime and magnesia of the Blackrock marl is much greater than any of the New Jersey beds, and the sand and silica are not in great excess. It really has as much fertilizing matter as the New Jersey marl, but it is deficient in the most valuable part, potash. This element, however, seems to be replaced by soda, which no doubt takes the place of potash in many vegetables where ash is rich in the alkalies.

§ 65. The sand of the marl beds of New Jersey varies from 39 to 70 per cent.; the remainder of which is more or less valuable in agriculture.

The phosphate of lime is probably the most variable in its quantity of all the valuable elements, and it is regarded as a mixture, and not forming a chemical union with either of its elements. Indeed it may in many specimens be seen and distinguished by its greenish gray color.

But it is never evenly distributed through the bed, as it has been ascertained by analysis, that it has occasionally accumulated in the inside of shells. It is, however, always present in the marl, and it no doubt exerts a favorable influence upon vegetables.

The upper bed at Blackrock differs in composition from the lower. It is less gritty to the touch, is of a darker green, more compact, and resembles a dark green clay. The sand in it is greater in quantity than in the lower, but is much finer.

On submitting it to analysis I found:

Sand or silex,	93.43
Peroxide of iron and alumina,	9.00
Carbonate of lime,	11.40
Magnesia,	0.20
Potash,	0.38
Soda,	0.42
Organic matter,	4.80
Water,	3.80
	100.43

The specimen submitted to analysis was taken near the upper part of the bed, about four feet above the line, along which the exogyra are the most numerous.

The results which I have finally obtained by the analysis of the green sand at Blackrock have disappointed me. I expected at least twice as much potash as I have been able to obtain; still when the green sand is carefully examined under the microscope it shows such a large intermixture of sand, and such imperfect green grains of the silicates, that would lead any one to expect on analysis unfavorable results.

The upper bed has, however, been tested as a fertilizer, and very excellent results have been obtained by its use.

The field immediately adjoining the bed of green sand had become so much exhausted that it produceed but three barrels of corn to the acre. Its employment the first year doubled the product of the field. The quantity employed was about two hundred bushels to the acre. The stalks of corn previous to its use were but little larger than the finger, and about half as long as the common growth in this latitude.

Previous to my last analysis of the marl of this locality I had hopes that it was sufficiently rich and valuable for transportation to the county of Chatham. If, however, on farther examination, beds can be found which contain from four to six per cent. of potash, there is no doubt it may be freighted in return boats to several points along the Deep river.

§ 66. The value of this species of marl is estimated from the amount of potash and phosphoric acid which it contains.

The price of marl in New Jersey is about eight cents per bushel. A bushel weighs, when it is wet from the bed, one hundred pounds. It loses, on drying in the atmosphere, twenty pounds.

The New Jersey fertilizer company deliver marl on board of vessels at their wharf for nine cents per bushel, and the white horse marl is delivered on the line of railroad, not exceeding ten miles from the beds or pits, for ninety cents. per ton. The potash in the different beds of New Jersey varies from two to seven per cent., very rarely as high as the last figure. At the pits individuals pay for marl from twenty-five to seventy-five cents per ton provided they perform the labor. The value of the potash in marl has been estimated at four cents per pound. Soluble phosphoric acid is estimated at five cents per pound, and the insoluble at two. But this distinction is uncalled for, inasmuch as all the phosphoric acid becomes available in time. The soluble, it is true, is more rapid in its effects, and produces more immediate results: it is no better for permanent improvements. Prof. Way, chemist to the royal agricultural society of England, has estimated the soluble phosphoric acid at eight and a half cents per pound, and the insoluble at three.

It must be recollected that in order to bring phosphoric acid to a soluble condition it requires considerable expense. It is better to purchase what is called the insoluble or tribasic phosphates than the soluble ones which are found in our markets and sold as superphosphate of lime.

The actual value of the mineral fertilizers to farmers is a question quite different from that which considers the value of bone dust, or potash by the pound. Immense benefits have been secured by the use of marl, which, considered in a commercial point of view, was worth nothing. The phosphoric acid in a bushel of shell marl is not worth, in commerce, a penny; but for use on worn out lands the farmer is enriched more than one-fourth of a dollar after paying for the labor of raising and applying it.

We are not, however, to confine our estimates of the value of a marl from its phosphoric acid and potash. Excluding the sand and insoluble silica, all the soluble matters are valuable to the farmer as fertilizers, and hence the determination of how much is soluble, and how much insoluble, is a more correct mode of getting at the value of marl than by confining our estimates to the two elements referred to.

These remarks apply only to the value of a marl for the private use of an individual owner, who employs his own hands in raising it when there is the least to do and economises his expenses to the best advantage.

Marl, however, in its crude state, as it exists in the pits, has a value which admits of estimation. The common shell marl may be hauled very frequently from two to four miles, and give profitable returns. This is often done. The shell marl, however, will not bear transportation as far as the green sand of Blackrock.

§ 67. I have alluded already to the difficulty of recognising certain marl beds in consequence in part of the absence of characters upon which geologists can rely. Among the beds of which there are doubts respecting their epoch, I find a green sandy deposit, which, if mineralogical characters may be relied upon, would be referred to the green sand which is now under consideration. They contain the green sand grains, but the characteristic fossils are absent except in one or two localities. The formation in question exists beneath the white or brownish shell marl at Mr. Flowers, Bladen county, Kingston, Lenoir county, on the Neuse, and at Tawboro', on the Tar river, and at many intermediate points on the banks of the creeks and ravines. It always occupies a position inferior to the shell marl, but as the latter are frequently absent, beds of sand and clay immediately succeed it. The green sandy beds at Mr. Flowers, beneath his shell marl, contain a few specimens of the Ostrea falcata, and at one or two of the bluffs above Mr. Flowers, on the Cape Fear, I found the vertebra of a large saurian, which I am confident belongs to the green sand, but in both of these cases their occurrence in these beds may have been accidental. I am inclined, however, in view of the few facts which bear upon the question of age, to refer these green sandy beds to the cretaceous system, occupying probably a position above these beds which have been described at Blackrock.

The predominent element of these beds is sand: if a sample is washed, a coarse sand remains, which amounts to two-thirds or three-fourths of the whole quantity employed. The quantity, in a few instances, may not exceed 60 per cent. Notwithstanding the large percentage of sand, it has been successfully employed as a fertilizer. I have, therefore, submitted several specimens to analysis, taken from different beds extending from the waters of the Cape Fear to the Tar.

A representation of the composition of this formation, as it exists at Mr. Flowers, in Bladen, and at Kinston, on the Neuse, is given in the following analysis.

§ 68. The Kinston green sand marl is of a dark green color in the bed, but becomes lighter when dry. Imperfect specimens of an Ostrea occur in it, but too much broken to be determined. It contains:

Sand,	91.000
Peroxide of iron and alumina,	4.700
Lime,	1.000
Magnesia,	0.700
Potash,	0.230
Soda,	0.260
Water,	1.500
Soluble silica,	0.204
	99.634

The marl, or this variety of green sand at Kingston, is one of the most sandy varieties known. It was regarded as too sandy to require the analysis to which it was submitted; but as the marl bed only one mile above had been successfully employed as a fertilizer, and appears to be equally charged with this useless element, I was desirous of knowing how this fact could be explained. It will be seen that the nine per cent. of fertilizing matter is really rich in potash, soda and lime, and, therefore, where a heavy dressing is applied, quite a large amount of this matter is added to the soil, and which contains a small quantity of potash. The sulphuric acid was not determined, but all of these beds contain it, which is no doubt derived from the sulphuret of iron or pyrites, which is always present.

An unfinished analysis of a parcel taken from a bed which occupies a similar geological position on the plantation of Col. Green, of Craven county, gave:

Silex or sand,	83.20
Peroxide of iron and alumina,	9.00
Lime,	2.31
Magnesia,	0.50
Water,	2.60

It lies beneath a white eocene marl, has a deep green color in the bed, but becomes brown after being exposed to the atmosphere. It has not been used as a fertilizer, but is undoubtedly richer than the Kingston marl which produces good effects upon corn.

A similar composition obtained in the same beds upon the

Tar river. A marl, for example, which has been used as a fertilizer by Hon. R. R. Bridges, contains:

Sand or silica,	89.700
Peroxide of iron and alumina,	5.000
Lime,	1.500
Magnesia,	0.200
Potash and soda,	0.250
Water,	3.510
	100.151

It is evident this variety of marl cannot be transported far because of its excess of sand, and in the instances in which it has been employed it has been transported only a short distance. These marls, however weak as they may appear, frequently destroy the existing vegetation. It is due to the existence of decomposing sulphuret of iron, which forms an astringent salt, copperas, or a mixture of sulphate of iron and alumina. This injurious salt is not formed where there is a sufficient quantity of lime to neutralize the salt, in which case gypsum will be formed. It should be remarked that the astringent salts may exert a beneficial influence where they are formed only in small quantities.

Another similar outcrop of this sand appears in the bed of a creek adjacent to the dwelling of Col. Clark, in Tawboro'. On submitting this marl to analysis I found it composed of

Sand,	91.300
Peroxide of iron and alumina,	5.800
Carbonate of lime,	0.190
Magnesia,	0.130
Potash,	0.150
Soda,	0.130
Sulphuric acid,	0.300
Water,	1.200
	99.200

A thin bed of the supposed upper part of the green sand formation appears in the series of beds on the banks of the Tar river, three miles from Tawboro'. At this bank the shell marl occurs in place, and has been used as a fertilizer by Col.

Clark with good success for many years; the relative position of this upper bed of green sand is represented in a section already described. It lies, as will be seen, immediately beneath the shell marl; and beneath the green sand a gray sand crops out, which is quite consolidated, and to the eye appears much like a limestone formation, but, as will appear in the sequel, is a bed of sand of unknown thickness.

The upper mass of green sand, which does not exceed four feet, has a similar composition to those already noticed. It is composed of

Sand,	79.000
Peroxide of iron and alumina,	8.800
Carbonate of lime,	2.752
Magnesia,	1.600
Potash,	1.739
Soda,	0.300
Soluble silica,	0.600
Sulphuric acid,	0.200
Organic matter,	2.000
Water,	2.330
	99.321

§ 69. Although the proportion of sand is large in this marl, yet I believe it is a more valuable fertilizer than the shell marl above it.

It contains more potash than the green sand of Black rock on the Cape Fear. It contains, it is true, less lime, but if the composition of the ash of the cotton stalk is consulted it will be perceived that magnesia is also required—this marl contains a large percentage of this substance.

It may be regarded as containing seventeen or eighteen per cent. of fertilizing matter. No trial has been made of this stratum, and of course nothing can be said upon the ground of trial.

§ 70. A very useless bed of gray sand occupies the bank at the water's edge, which has been alluded to. Nevertheless, I submitted a specimen of it to analysis. It is one of those beds which is charged with sulphuret of iron, and forms astringent salts, on decomposition, of the sulphuret of iron which is diffused through it. Beds of this description may be known by pouring muriatic acid over the material when a large quantity of sulphuretted hydrogen is liberated, which has the odor of rotten eggs—the smell of which is not usually forgotten.

This bed is composed of

Fine Sand,	93.500
Peroxide of iron and alumina,	2.000
Lime,	trace,
Magnesia,	trace,
Sulphuric acid,	1.000
Water,	3.200
Potash and soda, (undetermined,)	
	99.700

The bed is partially consolidated. It is, without doubt, entirely worthless as a fertilizer. As a geological formation it may probably be regarded as one of the beds of sand which separate the different beds composing the green sand proper; still, no opportunity has as yet been furnished me to see what lies beneath it.

The foregoing analyses of the green sand furnish all the necessary information respecting its composition. These beds in North-Carolina are deficient in potash, an element which, in New Jersey and Delaware, give to this fertilizer its importance. It is possible that exposures of other parts of this formation may come to light, which will be richer in potash. We do not obtain access to the best parts, which may be richer in this element. Other analyses, therefore, of new beds may result in better success, and finally furnish a fertilizer equally rich with those of New Jersey.

CHAPTER VIII.

Eocene or white marl.—Quantity or per centage of lime variable, but greater usually than in the other varieties.—The Wadsworth beds.—His letter and remarks.—Beds upon the Neuse.—Haughton's marl.—Composition, etc.

§ 71. In the ascending order, the next series of marls belong to that division of the formation which is known as tertiary, and that part of it which is called the eocene. This part is the oldest section of the division, and hence, reposes upon some part of the cretaceous system; either the green sand, which has been already considered, or else upon the chalk, as is the case in Europe.

Considered as a marl, it is readily distinguished from the green sand, even where its relations are concealed. The color is white, or else a light drab, or cream colored, and is very frequently made up of grains, which, when examined under the microscope, are found to be fragments of organic remains, such as corals, shells and echinoderms. Some beds, ten feet or more thick, are a mass of small fragments of fossils, mixed with sand. Some have a chalky whiteness, others take a brownish tinge. These beds are frequently soft, and may be loaded into a cart like dirt. In other cases, consolidation has taken place in part, and the mass is known as stone marl. This variety of marl is more calcareous than the green sand below, or the shell marl above, and when the mass is consolidated it makes a tolerable lime for agricultural purposes. But sand, which is a constant part of all formations in the eastern counties, exists in large proportions in some beds, and usually exceeds fifty per cent. But some beds have seventy or eighty per cent of lime, and when thus charged, the lime is well fitted for mortar, or whitewashing, as well as for agriculture.

§ 72. The eocene marl occupies a narrow but an ill-defined zone, stretching across several of the eastern counties, from the lower waters of the Cape Fear, in Hanover county, through a part of Onslow, Jones and Craven counties, crossing the Neuse twenty miles above Newbern, where it is either lost in the low grounds, or may be discontinued before it reaches Beaufort county, as the only marls of the lower waters of the Tar belong to the shell marl, or miocene beds; where the next bed below is visible, it is known to belong to the upper part of the green sand, which has been described.

The eocene is known to exist at Wilmington, at Pollocksville, in Jones county, and underlies the whole country in the vicinity of Newbern, upon the Neuse. In this formation I include the consolidated beds which have been employed for mill stones, and which consists of a mass of the casts of shells, the most common of which is a small species of clam. Recently, this variety has become an important building stone, and has been employed for enclosing the cemetery at Newbern, for which it is more suitable than any other rock which could have been procured.

§ 73. It will be seen from the foregoing remarks, that it occupies a less area than the green sand, and it will also prove to be more limited than the shell marl, though the latter never forms a continuous deposit over a large area. When in rocks, or consolidated, it is also broken up or traversed by fissures, and forms, if at the top of the ground, a very irregular surface.

§ 74. The white eocene marl has been used as a fertilizer, and probably with results as striking as the common shell marl. It would seem to possess some advantage over other marls, except the green sand, especially as it is fine and earthy. It is also richer in lime. For analysis I have selected several specimens from the central part of the region where it is underlaid with it.

The marl of Wm. Wadsworth, Esq., of Craven, furnishes a kind which represents its characteristics in as much perfection as any of the beds of the county. I found it composed of

Sand,	26.60
Water,	1.70
Magnesia,	0.10

Carbonate of lime, ... 71.22

99.62

The sand is in the form of white grains, often coarse. It is a soft, earthy marl, and is made up of fragments of corals, shells, crinoid's or pentacrinites, with sand mechanically mixed.

The influence of this marl upon vegetation has always been favorable, and the testimony of Mr. Wadsworth, whose ample experience qualifies him to advance an opinion, fully sustains the foregoing statement.

I subjoin an interesting letter from Mr. Wadsworth upon the subject of marl and marling. His observations, I have no doubt, will be concurred in by his neighbors. I am the more desirous of making his letter public on account of his experiment with marl upon his premises for the purpose of counteracting the tendency to fever and ague during the autumnal months. If farther trial should confirm the opinion expressed in favor of the use of marl as a preventive of fever, the importance of the discovery cannot be over-estimated:

CORE CREEK, CRAVEN COUNTY,
May 7th, 1857.

PROF. E. EMMONS—*Sir:*—The marl, (a specimen of which is sent,) I have been applying since 1852. I have now marled 220 acres. I have, until this year and a portion of the last, applied 100 bushels to the acre. I am now using 75. The weaker parts of my land were burned with the former quantity. My land varies from a very stiff clay to a soil quite light. Presuming you will be willing to be troubled with it, I will give you my mode of using it, and the results: My carts are made to hold just five bushels. I have the land checked off with the plough into as many squares to the acre as I design putting on bushels of marl. One bushel is put into each square. The first four bushels is pulled out with a hoe from the tail of the cart, and the last one is dumped.

By this method I am enabled to have the material much more equally spread, which I think is a full equivalent for the extra trouble. I usually begin to haul after my crop is "laid by," and it remains in the heaps until about the following February, when it is spread and ploughed in. I have spread some and let it lay on the surface twelve months before it was turned under, but I never saw any advantage from it. I have a small piece of very poor land that has been lying in that condition since the first of the year 1854. It was designed as an experiment. The growth on it when it was marled was altogether broom straw; there is now mixed with that growth some briars, dog fennel, and other weeds. I have consequently inferred there was some improvement, but whether it is as great as on land that was marled and cultivated I shall not know until I cultivate it.

The land I have marled and cultivated has very considerably improved. My whole crop has very nearly doubled, notwithstanding one-fifth of the land I crop on is yet unmarled.

I cultivated the land every other year in corn, and it rested the other, and not pastured. Last year I sowed peas on a portion of the rested land; what will be the result I am now unable to say. I have used plaster on the marled land, and have not seen any beneficial effect.

I fear I am trespassing too much on your time; I will, however, say a few words on my experience of the effects of liming on the health of the place. Before marl was used on this plantation it was uncommonly sickly, so much so that I was compelled to carry my family away every fall. Scarcely a person, white or black, escaped the ague and fever, if he had no more. All the land around the house has been marled, and the yard, under the houses, under and around the negro houses, I keep *freshly* marled every summer. Last summer I made my servants use it, as our grand mothers used to use sand, inside of the houses. Whether it is owing to this, or to a ditch I have had cut through the yard, or whether it is an accidental occurrence I can't say, but fall before last there was not a chill on the premises, and last fall there was but one case.

I will trouble you with one more *result:* These premises were infested with ants and fleas, now such animals are hardly known here.

W. B. WADSWORTH.

§ 75. In a subsequent letter Mr. Wadsworth's remarks go to confirm his previously expressed opinions, but that the reader may be benefitted by Mr. W.'s experience, I subjoin his remarks in his own language:

CRAVEN COUNTY, N. C., (NEAR NEWBERNE,)
October 12*th*, 1857.

PROF. E. EMMONS—*Dear Sir:*—The fever for marling is spreading in this part of our county and a good deal of land will be limed this winter. I have given some of mine an over dose with only one hundred bushels. Last fall and winter I used only seventy five and now I am putting on fifty. My experence so far has taught me to begin with a very limited quantity and to add to it as the land improves. Where I have not burned my land the improvement is very satisfactory.

I mentioned in my last letter to you the effect that marling, or ditching, or both combined, had had upon the health of this place. I told you that this plantation was remarkably sickly previous to the fall of 1855—so much so that it was strange for even one to escape billious, or ague and fever. I mentioned that in 1855 there was not a case of either, in 1856 but one, and now I will add that so far this fall, in a family of forty persons, there has been but two cases. (I happened to have been one of the subjects.) These three falls have been dry. I don't know how a wet one would act upon us. I have kept marl plentifully used in my yard, and around and in my negro houses.

I shall be under many obligations to you for analysis of my marl.

Yours, &c.,

W. B. WADSWORTH.

§ 76. A marl belonging to the same epoch, (eocene) furnished by J. H. Haughton, from his plantation in Jones county, gave me 56.06 per cent of carbonate of lime. Another specimen gave:

Silex or sand,	13.00
Phosphate of peroxide of iron and alumina,	1.10
Carbonate of lime,	85.20
Carbonate of magnesia,	1.02
Potash,	0.02
	100.34

I have found in these white marls a small per centage of potash. It is evidently less than in the other varieties. This is made up like the Wadsworth marl, of fragments of fossils, in which certain species of corals and a crinoid abound.

A variety is met with which is derived from the disintegration of a large species of oyster. It occurs upon the plantation now owned by L. Haughton, Esq., and is known as the Pollock place, in Jones county. It contains:

Carbonate of lime,	34.54
Sand,	63.46
Peroxide of iron and alumina,	1.30
	99.30

Large grains of sand are distributed through the marl. It

follows necessarily, from the manner in which these marls have accumulated, that they should vary in composition, and that the substanee which reduces the quantity of carbonate of lime, should be sand.

A ready method by which its quantity may be estimated is by washing a given quantity. It will be seen, that by agitating it in a vessel of water, there is a considerable quantity of fine, inpalpable white powder. Wash it until the water pours off clear, and the sand with the coarse fragments of fossils remain. The existence of much sand is not suspected at first, but as washing progresses, it will be found to prevail, in some cases, over the carbonate of lime.

§ 77. Upon the Neuse, about twenty miles above Newbern, heavy banks of the marl under notice occur, which extend continuously for more than a mile. This exposure of marl is upon the plantations of Samuel Biddle and Benjamin Biddle. It is accessible, and forms steep escarpments on the south side of the river. On account of the accessibility of this outcrop of marl, it will hereafter become an important deposit from the lime which it is capable of furnishing. It is consolidated, and may be quarried for the kiln, but it also furnishes an abundance of marl in a fine state of subdivision.

It has been tried imperfectly as a fertilizer, but while the result was disastrous, we may infer from it, that it possesses as valuable properties as the kind used by Mr. Wadsworth, which has been described already. The quantity used by Mr. Biddle, in his first experiment, was 600 bushels to the acre; consequently, most of the vegetation was killed, and very little has grown upon the land, thus excessively marled, for six years. It is just recovering from the dose. The consolidated part of this outcrop of marl contains:

Sand,	20.00
Carbonate of lime,	78.60
Oxide of iron and allumina,	1.70
	100.30

Another specimen of consolidated marl from Benjamin Biddle's plantation (Egypt) gave me:

Sand,	9.60
Peroxide of iron snd alumina, containing phosphoric acid,	4.40
Carbonate of lime,	85.00
Magnesia,	trace,
	99.00

A few grains of coarse sand were visible in the rock. This mass is evidently sufficiently pure for burning into lime. It would be adapted for the various purposes for which lime is required, as mortar, whitewashing, or for agriculture.

CHAPTER IX.

FERTILIZERS—CONTINUED.

Shell marl.—Heterogeneous in its composition, and arrangement of its materials.—Chemical constitution.—Application of marl.—Poisonous marl.—How corrected.—Theories respecting the operation of marl.

§ 78. The third bed of marl in the ascending order has been appropriately called *shell marl*, from the great abundance of undecomposed marine shells, of which it is mainly composed. The mass, taken as a whole, is formed of perfect shells, and those which have become fragments, and sand. There is no order in their arrangement in the bed. They lie as if they had been washed up on a beach; hence, they are mixed confusedly together. The relative position of the shell marl is exhibited in the sections already given. It is not present, however, even where all the other members of the sections in a bluff or outcrop exists. Whether its absence is due to denudation, or whether the beds were formed only at certain points, has not been determined. Denudation, however, has taken place at some of the beds, as they still preserve the gullies which were cut through them, and which were subsequently filled with brown earth.

Although it is not possible to detect an orderly arrangement of materials, still, certain parts occupy usually a common position; for instance, the large pebbles, coprolites, and certain bones and teeth lie at the bottom of the stratum. The inference which may be deduced from this fact is, that during the first stage of its formation, there was considerable violence in the movement of the waters in which the stratum was accumulating; and that probably, prior to, and during the early part of its accumulation, there were shiftings of the strata; some being more elevated, others depressed; or there was a change of level of the sea coast, which set in motion the waters, and led to the violence which collected at the bottom the large and less destructible fragments to which I have alluded.

But in the first place, I propose to speak of the use of this marl stratum as a fertilizer; and as it has a more general distribution, it has been employed more extensively than either of the foregoing which I have described.

The beds of shell marl are not composed uniformly of the same elements in the same proportions. It is as heterogeneous as possible in this respect. Some beds contain ninety per cent of sand; in others it is reduced to twenty-five per cent, and the remainder is mostly carbonate of lime.

§ 79. The most important subdivision which can be founded upon composition, is that into a gray or whitish marl in the mass, the color of which is due to the great abundance of marine shells, and that of a dark bluish green marl, which contains grains of green sand. In the latter there is a notable amount of potash, while in the former it exists only in very small proportions. Some recognize a red or brown marl. This color, however, is due merely to exposure to the atmosphere, in consequence of which the protoxide of iron has changed, or is changing, by the absorption of oxygen into the peroxide. This change is indicative of a valuable marl, but it is no better subsequent to this change than before it. If in the greenish marl green grains can be distinguished, it may be inferred that the marl contains potash. The presence of carbonate of lime, as is usually known, is indicated by effervescence when acids are poured over it, and a judgment may be formed by its continuance and violence, whether it is rich in this substance. If it is prolonged, there is a large quantity of carbonate of lime in the specimen under examination. So the presence of sand may be detected and its quantity proximately determined by simple washing.

§ 80. The shell marl upon the Cape Fear river belongs usually to the former. A bed, however, in the bluff at Brown's landing, contains the green grains alluded to, but still it is readily distinguished from that upon the Tar river, which is usually bluish green, and belongs to the latter variety. I do not, however, attach much importance to the subdivision.

There are several beds of shell marl immediately upon the banks of the Cape Fear, or within a mile of them; and when marine shells are closely packed in the strata their several compositions are alike. As a representation of the composition of this marl, I shall select Mr. Cromarty's marl bed, near Elizabethtown. It consists mainly of:

Sand,	52.50
Carbonate of lime,	40.25
Peroxide of iron and alumina,	7.20
Magnesia,	0.75
Potash and soda,	traces.

I have always found phosphoric acid when the peroxide of iron and alumina are tested with molybdate of ammonia. It is very rare for the carbonate of lime to amount to seventy-five per cent. I found seventy-one per cent in Mr. McDaniel's marl, in Nash county. The bluish green marl of Tar river is quite sandy, and yet may be regarded as a rich marl. As an illustration of this fact, I subjoin an analysis of the marl bed owned by Col. Clark, three miles above Tawboro', on the Tar river. It consists of:

Peroxide of iron and alumina,	6.80
Carbonate of lime,	16.10
Magnesia,	0.436
Potash,	0.616
Soda,	1.988
Sulphuric acid,	0.200
Soluble silica,	0.440
Chlorine,	0.030
Phosphorie acid,	0.200
Sand,	72.600

Of one hundred parts, only about twenty-six can be regarded as available matter, and yet good results have attended its use.

Immediately above the shell marl of the Tar there is a bed of clay some four feet thick. This clay I have submitted to analysis for the purpose of ascertaining the quantity of potash it contains. The results show, however, that as a fertilizer, it is of no importance. It gave me:

Sand,	84.00
Peroxide of iron and alumina,	4.40
Lime,	0.35
Magnesia,	0.10
Potash,	0.05
Soda,	0.02
Soluble silica,	0.20
Organic matter and water,	10.50
	99.62

All the beds except the upper beds of sand were submitted to analysis. Only two in this bank are valuable fertilizers, the shell marl and the upper bed of green sand; both contain potash, soda and phosphoric acid; and there is no necessity for rejecting the latter when hauling marl for the plantation. If some method could be devised by which the sand could be cheaply separated from the mass, the remainder would form a marl superior to the richest green sand; the sand being coarse, presents a favorable condition for effecting a separation.

§ 81. The green shell marl of Mr. Bridger's plantation, upon Fishing creek, I found to possess a composition similar to Col. Clark's. There is a greater proportion of sand, but the available part is almost identical with the Tar river marl.

§ 82. The application of marl is an important matter, and requires a brief discussion. Notwithstanding marl has been used for many years, still there is much disagreement among planters of experience as to the best mode of applying it, and the quantity to be applied in any given case. Its effects are frequently deleterious if a large quantity is spread upon a poor soil, and yet it has not been ascertained how its injurious effects may be obviated. It is no doubt desirable in many instances to use a larger quantity of marl than the soil will admit of when it is in its natural state.

The quantity of marl which is usually spread upon an acre of ground is from 150 to 200 bushels. Three hundred bushels is often used. But certain worn out lands would be exceedingly injured for several years by even two hundred bushels. The question, I have no doubt, has been often put: Why is marl ever injurious? The natural conclusion is that it contains some substance unfriendly to vegetation. This substance is no doubt in certain cases an astringent salt, formed in those marls which contain iron pyrites which is prone to decompose on exposure to those bodies which contain oxygen, the sulphur thereby is oxidated, and slowly acts upon the iron and forms copperas, or upon alumina, which is present in the marl. In small doses copperas will not fatally injure vegetation, but operates beneficially. The term in common use for expressing the effect of injurious marls is, *burning*. Those which are decidedly burning marls have the distinct taste of copperas, sometimes it appears upon the surface of those marls in dry weather, when it has a whitish appearance. But gypsum sometimes appears also. This may be distinguished from copperas by being tasteless.

§ 83. There is no difficulty in treating marls in which copperas is found. It is readily decomposed by lime. Let a compost heap containing a hundred bushels of marl be formed, mixing leaves or any organic matter as stable manure, and then add three bushels of quick lime to the mass, and incorporate the ingredients together by shoveling them over twice. Gypsum will be formed by combining with the sulphuric acid in combination with the iron. The compost is all the better for the lime, though it is possible the gypsum may not in all instances prove itself useful. Astringent marls, when in heaps in the open air, lose their copperas and other soluble salts by solution in rain water to which they are necessarily exposed, they undergo a leaching process by which they are freed of their injurious properties. Another method may be resorted to when it is found that vegetation is being injured, or has been by the experience during the year of its application, to plough deep and mix the marl with a large quantity of soil; the fertility will be restored. It is by no means difficult for any farmer to test his marl prior to its use if he wishes to ascertain whether this astringent salt is present. To do this, let the marl be boiled in rain water; strain it, or let the turbidness of the solution disappear by rest; pour off the clear liquid, and if sulphate of iron and alumina is present, it will turn black by adding a solution of strong tea to it; it will become a dirty white by lime water and a solution of the leaves of red cabbage change it to red, showing the presence of an acid salt. Most of the marls of the State contain these salts. Where they are abundant undecomposed pyrities will be found in masses adhering to portions of petrified wood or inseparate concretions in the marl.

§ 84. Writers upon the efficacy of marl as a fertilizer, have entertained different opinions. As the progress of agriculture has been promoted, and observation and experiments multiplied upon the effects of different bodies upon vegetation, these opinions have become more consistent and reliable.

Some writers have maintained that lime alone is the effective agent; others that it is pyrites, or else is due to the presence of animal matter, which has been derived from the fossils of the beds; others, still, to the presence of phosphate of lime, while others have maintained that it is due to the potash.

§ 85. Now, it is quite possible that all these opinions are right as far as they go. They are erroneous in being restrictive. If we examine the composition of an ash of any plant, as I have already observed, we shall find all these elements, and we may well suppose, as they are all so generally present, that they are all required; and hence, we are not to attribute the efficacy of marl to one of its elements exclusive of the others. It may be, that a given soil is notably deficient in potash, while the other elements are in sufficient abundance to furnish all that a given plant requires. In such a case it might appear that fertility was restored to the soil by potash alone. Of all fertilizers, wood ashes are the best, and possess a more general application than any other; being adapted to any crop. They are the best, because they contain all the elements the plant needs; and hence, the nearer a marl is in composition to wood ashes, the better it is. Hence, then, the efficacy of marl is due to its potash, soda, lime, iron, magnesia, phosphoric acid, sulphuric acid and chlorine, and not any one of its elements, exclusive of the others. The only modification which this doctrine requires, is that some of the elements are more important than others, and it may be true, that the controlling influence is to be ascribed to the alkalies, alkaline earths and phosphates; still, the marl is better with the less essential elements, than it would be without them. The absolute value of a marl is shown: 1., by the amount of soluble matter it contains. 2., by the predominance of the most valuable elements, as potash and phosphoric acid. Marls which contain the most of these bodies are the quickest and the most durable in their effects; and when the marl is rich in them, a full dressing lasts from fifteen to twenty years.

§ 86. In forming a theory respecting the active elements in marl, our views should not be limited to the nutrient properties they possess, or simply to the *food elements* which contribute directly something to the weight or growth of the plant.

Some elements perform a function in growth or nutrition, which is independent of nutrition in this sense, or they are *nutritive* from their reactive forces; they are not taken up by the plant, but furnish or provide a substance by their reactions upon each other, which is nutritive or administers to its growth.

These substances perform a double function; they are really nutriments, and are taken up into the vegetable tissue; but, in addition to this, their reactions upon other matters in the soil are such that nutrient matter is constantly provided without their increase or diminution in the soil or marl.

The substances which are known to perform a double office, are the oxides of iron and organic matters. To enable me to give a brief exposition of the functions of the oxides of iron, I will state what takes place in the soil when it is well constituted for the growth of cereals, and other plants employed as food. It will be observed that in the analysis of soils, the iron is set down as a peroxide; this is the state in which the iron is obtained. In the best of soils the iron is not all of it in this state; but that of a mixture of the two oxides—the protoxide and peroxide. Now, the protoxide is changed in making an analysis into the peroxide, by the addition of a few drops of nitric to the hydrochloric acid, which is employed for effecting a solution, for the purpose of obtaining an exact or an uniform result. The nitric acid added to the solution, is deprived of so much of its oxygen by the protoxide as is sufficient to change it, or convert it to a peroxide. Now, in the ordinary course of nature, this change takes place when the soil is freely exposed to the action of water and air. The protoxide passes into a peroxide by the absorption of oxygen from the water. It would remain in this state permanently, if the soil was dry and free from vegetable or organic matter. When soils become exhausted of these matters, it remains a permanent peroxide. If, however, this peroxide comes in contact with organic matter, it robs the peroxide of an equivalent of oxygen, and passes again into the condition of a protoxide. It is possible, therefore, for these changes to take place at all times when the needful conditions exist. But this is not all; the water of the soil being robbed of its oxygen, its hydrogen is set free; and being in its nascent state, it is ready itself to combine with that body, for which it has the strongest affinity. That body is *nitrogen* contained in the air diffused in the soil; and the body formed by this union is *ammonia*. Now, ammonia is one of the most essential bodies in the list of nutrients. Guano, as is well known, owes its fertilizing properties in part to ammonia. But I need not dwell upon this fact. By the interchanges of oxygen which take place with the oxides of iron, we are furnished with an explanation of the origin of ammonia in the soil. But the production of ammonia is only one of the chemical changes which take place in a soil in which organic matter, iron, water and air exists. The vegetable matter, also, undergoes a change, for the oxygen which it has taken from the peroxide of iron converts it into organic acids, which are known by the names of *crenic* and *apocrenic* acids. These acids being one of the series of changes effected through the influence of the oxides, they in their turn become active, and unite with the ammonia and form crenates and apocrenates of ammonia. In the condition of a salt, this compound of ammonia and the vegetable acids are taken up by the roots of plants, and become their food.

§ 87. I have made these remarks for the purpose of preparing the way for farther observations upon the action of marls upon vegetation. The condition of the iron in a large proportion of the marls, is that of a protoxide. Thus the iron in the greenish marl upon the Tar River, is a protoxide. In this condition, when it is spread upon land and mixed with the soil which contains vegetable or organic matter, changes first into a peroxide, it is then in an active state, and sezing upon one of the elements of water, decomposes it. The oxides of iron in the marl undergo the same changes in the soil to which they are applied, as those which have been described as taking place in all soils which have not been exhausted of these organic matters. It will therefore be expected that marls which contain a large percentage of iron, are more valuable than those which are destitute of it, and to the action of its oxides, we are indebted for one of its most important effects, the supply of the salts of ammonia, and even the organic salts of potash, soda, and lime.

These facts furnish important hints relative to the proper preparation of marl for the plantation, viz: that it should be composted with organic matters. We supply in this way the conditions for its favorable action upon vegetation. With a large quantity of organic matter, a large amount of marl may be used without detriment to the vegetation, and the larger the quantity the greater the amount of ammonia which will be generated. For certain crops, this practice is of the highest importance. It has been proved by numerous experiments with wheat, that there is a certain yield produced by the use of the mineral fertilizers as phosphates of lime,—but these will not increase the yield beyond a certain standard when used by themselves. But if a larger supply of ammonia is furnished, the number of bushels per acre is increased beyond that standard. So that in order to bring lands to their full capacity, ammonia must be supplied also directly, or indirectly. A compost of marl properly made, is one of the best fertilizers for wheat, and there is little doubt, that the favorable influence is due in part, to the chemical changes which I have described by which ammonia is one of the products of change.

To estimate, therefore, the value of marl by the number of pounds of phophoric acid and potash which is contained in a ton, does not give its true value. All marl contains a small amount of organic matter, but it is improved by adding more, and thus preparing it, we provide for a continuance of those changes by the instrumentality of iron until the organic matter is consumed, and when ammonia will cease to be generated. It will be understood, therefore, that organic matter is necessary to effect these changes which produce the salts of ammonia; in its total absence, it is true, ammonia is produced; still, in the state of simple ammonia, it is not fit for nutrition; it requires a union with some acid, and therefore the great importance of providing all the conditions for the full action of marl upon the crops to which it is applied.

§ 88. If the foregoing views are correct, it will be admitted that the simple application of the oxides of iron and organic matter may become the best of fertilizers. Experience has proved that the scales of black oxide of iron, or the oxides and other refuse matter obtained from a smith's forge are excellent fertilizers for the pear and other fruit trees; and they are no doubt equally valuable for wheat and Indian corn. Iron itself is always present in the ash of a plant. It is no doubt an important element in its organization, giving it tone and strength. But as we have attempted to explain, it is equally an essential element in soils and marls, for its influence in effecting those changes which finally result in the production of the vegetable salts of ammonia, potash, soda and lime. It is in this state that they are taken up by the roots of plants and become thereby the effective agents of growth.

When the functions of iron in a soil or marl are known, it does not appear improbable that it is as important and as valuable as phosphoric acid or potash. In some marls it is easy to recognise the change which the iron has already undergone by its having become brown or reddish. This change does not probably affect its qualities, though some maintain that the red marl is better than the blue. The only difference between them is, that the protoxide of the blue has passed into peroxide; the latter may be changed back to the protoxide in a soil charged with organic matter, and though I have omitted to state the fact, the organic acids are capable of acting also upon the oxide of iron and forming with them salts, in which state they become fitted for reception into the circulation of the plant.

§ 89. I have dwelt somowhat at length upon the importance of the oxides of iron and organic matter in the soil. This subject is especially interesting to planters in this State, 1st, from the fact that so large a proportion of the best soils of the eastern counties consist of vegetable matter in the main, and 2d, from another fact that the soil in the midland counties is deficient in organic matter, it having been consumed by long cultivation, aided, in a considerable degree, by climate. In 1847, I prepared an article for the American Journal of Science and Agriculture, the object of which was to set forth in as a clear a light as possible, the functions of the vegetable matter in the soil, and having seen no reason for changing the views I then entertained, and still believing them to contain important principles, I shall transcribe them as they were then printed. It should be remarked, however, that the more scientific details of the paper belong to the celebrated Mulder, who has taken a widely different view of the importance of organic matter in soil from Liebig. I made just an allusion to the doctrines inculcated in a previous communication, which is contained in the following extract:

"Supplying, then, the soil with decomposing organic matter, and several important results follow; the rocks are dissolved and the plants may be supplied with the necessary carbon, ammonia, and other essential inorganic matter." The doctrine contained in this extract is important, and may be drawn out more in detail. The opinion has generally prevailed that mould or the black matter of soil, was eminently useful. Many, and perhaps all, at one time entertained the idea that it was the principal food of plants. The idea, it is true, was crude, and it will not offend any one at the present time to say that the early notions of farmers and chemists, who had turned their attention to the subject, were crude, and probably, if we insist upon it, were really erroneous. Still, even error, in toto, is rare, and some truth at least is usually mixed with it; that it was a valuable composition in the soil, and performed some function serviceable to vegetation, was a common belief. The error consisted in the misapprehension of the truth, and was not so broad or fatal as that which maintains that it is of no use at all. It is by no means a fatal error to maintain that a substance is important, and yet mistake its function or office. It is one of those errors which belong to theory, and does not necessarily exist in practice. A farmer, for instance, believes that barn yard manure is useful. His belief will lead him to save it, and employ it upon his corn, and this he may do notwithstanding his theory of its action is misapprehened, or may be totally false. The main thing is to be right as to the fact. Still, a correct view of the whole subject, how the organic matter acts, in what way it is beneficial, and how it is related to the inorganic matter, will undoubtedly increase our power over the products of the earth. This is by no means an irrational view of the subject. If we apply it to some of the most common processes of farming, as plowing, it is evident that the farmer who best understands the object and use of plowing, will derive the most benefit from it. All agree that it is useful, and hence all will plow; still, those will plow the best, and adapt the work better to the end in view, who best understands its use, than the farmer who has only this naked truth at his elbow, that it is useful, but knows not why or wherefore. Theory, then, to continue the line of remark, is useful; and correct theory eminently useful. At the same time, the *fact* may, and usually is, more important practically; for the fact leads to the right action, but it may fall short of the benefit it is calculated to give, when fact and correct theory are conjoined, and go to the work together. Theory and book learning are often ridiculed by the *matter of fact man*, and yet observation often bears us out in the opinion that in most instances there is not only a great want of facts, but that also when found they are often greatly perverted. But we turn now to the subject more immediately before us. What are the functions which the organic matter performs in vegetation? Our belief is, that all terrestrial plants, if they do not absolutely require it, are at least benefited by it. That it is not taken into the plant in the condition of mould or humus, is proved from the fact that it is not in this condition sufficiently soluble. If then it is useful, it is necessary to maintain that it undergoes certain changes before it becomes the food of plants. It may minister to the wants of vegetation in several ways, without its becoming the food itself. It ministers to the vegetable by its presence, procuring thereby an open state of the soil, by which air is more freely conveyed to the roots. It ministers, also, to the wants of vegetation by its absorbent and retentive powers. Indeed, in this respect it is almost indispensable to vegetation. These, then, though not all the uses which mould exercises in vegetation, still are sufficiently important to merit the attention of the agriculturist. In neither do we find that the brown or black matter of soil becomes the nutriment of vegetables, and yet its service is immense. To understand the nature of the changes which take place in the organic matter of the soil, it is necessary to know what agents exist there. A mixture of carbonate of lime and magnesia, silex and alumine, and organic matter, would remain without change forever, were there no other bodies of a more active kind, whose affinities become a present and efficient cause for action. These powers or forces exist in the atmosphere and in the water diffused through the soil, and it is proper to make a distinction of the atmosphere within the soil, from that above or without it. The atmosphere is composed of two elements, oxygen and nitrogen, in the proportion of 79 nitrogen to 21 oxygen. The latter is free and uncombined with the nitrogen, or is merely dissolved in it, just as sugar or salt is dissolved in water. The consequences which follow from this condition or state of the elements, is, that both are free to unite with other bodies, that is, so far as attraction for each other is concerned there is no hindrance or force to be overcome to bring about a separation. Hence, in the respiration of animals, the oxygen of the atmosphere which is inhaled combines readily with the carbon suspended in the return or venous blood. So in the soil, there is the same independence; the oxygen or nitrogen is not hindered from uniting with other bodies by any affinity existing between themselves. The final end or cause of this is, the ultimate union of the oxygen with certain bodies in the soil, especially with the organic part. The other agent, water, undergoes chemical changes of a different kind. In this the elements are chemically combined, and hence they are not so readily separated from each other, and hence, too, its action is constant, and that which is proper to it in its state of integrity—it is the solvent power so necessary to bring all particles to a state of fineness that they may pass into the organism of vegetables; for solution is merely that separation of particles to that degree of minuteness that they are capable of being suspended in the medium. They are merely farther apart, and they are brought thereby into a condition to undergo farther and more thorough changes than they were previous to their solution or suspension in the medium itself. But certain bodies can and do decompose it, the final end or cause of which is to supply ammonia or rather nitrogen to the growing plants. Air and water, then, contain the elements which make it possible for the organic matter of the soil to return once more to that vital state in which it exists in living vegetables, or in other words, to become the food of plants.

If we now trace the changes which decaying wood undergoes from the time when it first ceases to be a living body to that last change by which it is fitted for the function of nutrition, we shall be able to see its use in this part of the economy of nature. Wood, when it has lost its vitality, goes to decay, but the progressive changes which it passes through are not analagous to putrefaction. Rotten wood, as it exists in decayed trees, is a neutral substance; neither acid or alkaline at first. But in progress of time, several definite substances are formed from it, which possess activity and belong mainly to the class of acids, and are capable of combining with the alkalies and alkaline earths which are soluble salts, and in this state minister to the growth of plants. Of the substances which are formed by decaying wood, and by peat or muck, ulmine is one, which is also a neutral body, and is quite insoluble, and hence is not useful as a nutriment. This substance is called ulmine from the fact that it was first prepared from the wood of the elm; but it is found in all other kinds of vegetable matters which are undergoing the changes already alluded to. Ulmine is formed from wood, or fibrous, vegetable matter of any kind, as leaves, twigs, &c., by the absorption of oxygen from the air, or contained in the moist earth. By a simultaneous action carbonic acid is liberated. The substance formed may be represented by C_{33}, H_{27}, O_{24}; 33 equivalents of carbon, 27 of hydrogen, and 24 of oxygen. The substance represented by this formula is a white, friable substance, found in the interior of hollow, decaying trees, and is produced by the oxidation of the woody fibre. Lignine also produces other bodies by combining with oxygen. Thus, 4 atoms of lignine,* C_{48}, H_{32}, O_{32}, with 14 of oxygen, produce 8C. O_2 with 18H. O.; and an atom of ulmine, C_{40}, H_{14}, O_{12}. Other products of an analogous kind are formed from wood by union with oxygen. Of these, humus and humic acids are

* Kane's Chemistry, edited by Draper, p. 638.

among the most remarkable. The first is represented by the formula C_{40}, H_{14}, O_{12}; the latter by C_{40}, H_{15}, O_{15}. These two acids, which are spontaneously formed, and are common in peat and other earths, differ from each other in their relations to ammonia; the first having no affinity for it, while in the latter it is so strong that it is difficult to separate them. In consequence of this affinity, it no doubt forms an important element in productive soils.

Another class of vegetable acids, which are also produced by the action of oxygen on organic matter, is called the azotized, from the fact that they contain nitrogen. These acids are the *crenic* and *apocrenic* of Berzelius. Both are soluble in water and alcohol; the apocrenic less so than the crenic. They form with alkalies and alkaline earths, soluble and insoluble salts; some of which are essential constituents of a rich and productive soil.

By the continued absorption of oxygen from the atmosphere, wood and other organic matters are converted into a nutriment for vegetables. The crenic and apocrenic acids are products of bodies which are nitrogenous themselves; the nitrogen of which is retained through all the changes which the organic matters pass.

It seems to be established, then, that organic matter may be useful to plants, and may promote their growth in various ways. This conclusion might be made almost *a priori*, subsequent to the determination of the nature of the bodies under consideration; for it is well known that many bodies require nitrogen; and it is ascertained that some of the organic bodies contain, and others absorb and retain ammonia obstinately. And each of these classes of bodies are soluble, and in a condition to be received into the vegetable system.

If the foregoing considerations are true, why should farmers be taught that the organic matter of decaying leaves and of their barn yards is useless? that it is a bad economy to spread it upon their fields, or plow it into their soil? We have sometimes wondered why it is that many intelligent farmers hold book farming in such low repute. We, however, have been satisfied as to the cause; when, for instance, doctrines are taught so contrary to their experience; and when they are told that they had better burn their barn yard manure rather than carry it out to their meadows, we are not at all surprised that they lose confidence in books, and hence often refuse to receive many things which are really sound and valuable; and this, on account of the erroneous doctrines which come apparently from a responsible source.

But to return to the consideration of ammonia in the soil. Chemists are not agreed as to the processes by which ammonia is supplied to the soil. That it exists there, and that it is provided for by certain chemical changes is admitted. We have stated in a former article in this journal, that one of the means by which it is restored to the soil is through the mutual influence of water and the protoxide of iron; the latter substance having the power of decomposing the former and taking to itself its oxygen; the hydrogen being liberated instantly combines with the nitrogen of the air in the soil, and forms with it ammonia. Humic acid, too, by its strong affinity for ammonia, rapidly absorbs it whenever it is freed from its combinations. Other modes undoubtedly exist by which the nitrogenous compounds are supplied with this essential element. Ammonia, too, has been proved to be present at all times in the atmosphere, though only in small proportions.

One of the forms in which ammonia is found in the soil is that of apocrenate of ammonia; a compound which is formed from humic acid by its continued oxidation; the apocrenic acid being merely a higher state of oxidation of the same substance. In the chain of causes by which apocrenic acid is formed, nitric acid is also generated, according to Mulder—this acid acts with great vehemence upon humic acid. Admitting the fact of the formation of nitric acid, and its subsequent action on humic acid follows necessarily; and furthermore, we can understand how the humic acid is oxidated and changed into apocrenic acid. Mulder says, p. 166, in his Chemistry of Vegetable and Animal Physiology, when apocrenic acid is found in the soil it is accompanied with the production of carbonic acid; the ammonia of the soil produced in it from the atmospheric air it has absorbed, may, by the influence of decaying, organic substances and water, be converted into nitric acid; and no doubt is so when the bases required for nitrification are present. Saltpetre was long extracted from the soil exclusively, as in many places in Egypt, India, &c. By the oxygen of the atmospheric air contained in the soil, the hydrogen and nitrogen of ammonia produced from the constituents of the air are oxidized; water and nitric acid as soon as it is formed, meets with a substance in the soil, humic acid and humin, which by its influence is converted into apocrenate of ammonia, and at the same time produces carbonic acid. This change of humic acid into apocrenic acid takes place in minute quantities; as is the case with the formation of ammonia which precedes it. Thus, to form one equivalent of apocrenic acid, there are required two equivalents of humic acid and one equivalent of ammonia and seventy-six equivalents of oxygen. In this production of apocrenic acid, the ammonia from the humate of ammonia is not only transferred to the apocrenic acid, but it performs an intermediate part, namely, the fixing of oxygen. Through the tendency of ammonia to form nitric acid, the oxygen of the atmospheric air contained in the soil is combined with the constituents of the humic acid; the ammonia itself remaining unchanged; neither leaving the soil, nor being oxidized into nitric acid. If there be not an abundance of organic matter, and if the air be moist, and lime, magnesia or potash be present, ammonia is first produced, and afterwards nitric acid. If, on the contrary, instead of these leaves, organic substances are in excess, humic acid is formed by their decay; at the same time, ammonia is produced from the nitrogen of the atmosphere; and, finally, apocrenate of ammonia, carbonic acid and water."

This long extract seemed to be required in order to put the reader in possession of the views of Mulder on this important subject; from which it is well established that organic matter in soil is of the highest moment; and that it not only ministers indirectly to the growth of plants, as stated in the early part of this article, but also becomes food itself in the form of apocrenate of ammonia. So, also, that important substance, carbonic acid, is liberated and furnished to the roots; a substance which many suppose is taken up by the leaves only. The apocrenates are continually forming; not only the apocrenate of ammonia but also those of potash, lime and magnesia.

Through, then, the action of the organic acids the inorganic bodies are received also into the circulation of vegetables; and this gives us an idea of its importance, namely, as a medium by which lime, magnesia and potash are supplied to the vegetable kingdom. The carbonates of lime and magnesia are rather insoluble bodies, though the carbonate of soda and potash are, as is well known, highly soluble.

We should take an unsafe course in practice, then, in rejecting the organic part of manures; and how truly important lime, potash, soda, magnesia, &c., are; still, soils cannot be and are not fertile if they contain only these; and the highest and most valuable soils are those in which a due balance is preserved between the organic and the inorganic parts.

§ 90. Unfortunately for the best interests of agriculture, the marls of North Carolina are too sandy to bear transportation to distant points; and hence, their use is now limited to the plantations upon which they are found. If, however, a method could be devised by which the sand could be separated cheaply from their useful parts, they would then be reduced in weight and bulk sufficiently to bear transportation on those railroads which pass within three or four miles of the beds in which they lie, and those especially upon the Cape Fear and the Neuse might be transported very cheaply by water. The quantity of sand, it will be perceived, is often as high as 80 per cent. The remainder twenty per cent contains all the fertilizing matter. This 20 per cent is a concentrated manure, and compares very favorably with the superphosphate of lime, especially, considering that its cost would be very much less, or according to its actual cost, it would be worth quite as much as the superphosphate.

By aid of suitable machinery, it is highly probable the sand may be separated rapidly from the valuable parts which compose it. If so, the interests of agriculture would be greatly promoted, and the revenue upon the railroads increased; and in the end, it might, and invariably would supplant guano, which is a drain upon the pockets of planters.

§ 91. In order to free the sand from adherent marl, it might be passed through a cylinder, the inside of which had many projecting angles, and within which another cylinder studded with angular rods should be made to revolve rapidly, while the marl and water was passing through them. The sand, after issuing from the machine, would subside almost immediately, while the lighter marl would pass forward and be allowed to subside in vats. With a machine properly constructed, a hundred tons of marl might be washed in a day, and though all the sand might not be removed from it, yet a very large proportion would be. Some of the marls, as analysis proves, contain seventy-five per cent of sand. The concentration consequent upon its removal would convert it into a fertilizer which would contain three or four times its amount if it was in its natural state. The washed marl would then possess the following composition:

Phosphate of lime,	2.50
Peroxide of iron and alumina,	25.00
Carbonate of lime,	44.17
Magnesia,	1.71
Potash,	2.35
Soda,	2.50
Sulphuric acid,	0.72
Chlorine,	0.52
Organic matter,	16.12
Soluble silica,	0.78
Water,	3.75

The commercial value of marl of this description will be from 8 to 9 cents per bushel. A bushel of dry marl weighing eighty pounds, and twenty-five bushels weighing two thousand pounds, it will be worth from $1 60 to $1 80 per ton. Fifty tons of marl might be washed per day, which would give about twelve tons of concentrated marl in the vats. The cost of raising and washing may be performed at from 37½ to 50 cents per ton, and perhaps less than the lowest figure.

§ 92. The washing of the marls should not be confined to the green sand marls, the white eocene marls upon the Neuse in Craven county, may also be profitably subjected to the operation. It would at any rate improve it much, for agriculture, and serve to create a demand for it in the midland counties. Besides, when it has been subjected to this operation, it becomes an excellent material for burning into quick lime. Being in a fine incoherent state after washing, and also wet or a calcareous mud, it might be pressed at once by means of moulds into the form of large bricks, and when allowed to dry, put up in kilns for burning. In western New York, the white fresh water marl is treated in this way, with the exception that it does not require washing. But it is moulded into the form of bricks and burned. It is highly esteemed for its whiteness, and is used mostly for white-washing.

The foregoing hints are thrown out without having had time and opportunity for testing their value. They are suggested in consequence of the scarcity of limestone in the middle counties of the State, and the consequent high price of lime. There is lime enough in the eastern counties, but its intermixture with sand, which diminishes its value in a commercial point of view, except in the case of a few banks, which have been designated.

§ 93. To show that green sand and other marls may be transported over railroads, I propose to quote what has transpired already in New Jersey,* thus, there was transported over the Freehold and Jamesburg Agricultural Railroad during 1856, 270,982 bushels of marl, all of which found a market out of the marl district, and some of it out of the State; and as an evidence of the estimation of the marl and the ready sale it finds along the road, it requires only to witness the high cultivation of the lands along the whole route of the road. Monmouth county, and other parts of New Jersey, were as barren, or as much exhausted by cultivation, as any

* Third Annual Report of the Geol. Survey of the State of New Jersey, for the year 1856, p. 58.

parts of this State. The use of marl has renovated the country, a profitable trade has sprung up which will not only benefit the owners of marl pits, but that part of the agricultural community who avail themselves of this substance, when it can be brought from a distance to their doors.

§ 94. The mode of calculating the money value of a marl, is founded upon the fact, that the percentages represent the absolute weights in the compound,—thus one per cent. of phosphate of lime is equivalent to one pound in a hundred. This number, one, or one pound multiplied by 20, and then estimated by the value per pound of the substance, gives its value in 10C lbs. of marl; or, if there is 2,16 phosphoric acid, the product is 4,32, which multiplied by 5 cents, the value per pound of phosphoric acid gives $2.16,0, or two dollars and sixteen cents, the value of this substance in a hundred pounds of marl. The object to be secured in washing the marl, is to raise the percentage of phosphoric acid sufficiently to make it a merchantable substance, and thereby benefit the agricultural community far and wide.

CHAPTER V.

Animal manures—Fish—Crabs—Cancerine composition of fish before and after drying—Compost of Crabs—Preservation of the offal of fish at the large fishing establishments.

§ 95. The best interests of agriculture require a ready and cheap supply of manure. Its prosperity depends upon it. Without fertilizers, it would be impracticable to sustain this branch of business, except in some highly favored districts where the supply has been prodigally provided. A source from whence an immense supply in some localities may be obtained is the ocean. The myriads of fish, for example, which resort to the shores of North Carolina, might be turned to an immense profit. The use of fish, employed for this purpose, has been practiced for a century upon and near the coast where they can be readily procured. Both Connecticut and Massachusetts have experienced the benefit of their employment. Recently in New Jersey a more systematic attempt has been made to furnish agriculturists with a supply of this kind of manure. In the old way of employing fish they were put whole, if small, into a hill of corn or spread over the field. In this mode they become highly useful, but were very offensive. The moss-bonkers have been principally used in New Jersey, and are regarded as a powerful manure. Prof. Cook has given an analysis of this fish for the purpose of ascertaining the amount of fertilizing matter which it contains and its comparative value when dried as a manure.*

In the fresh state, it consists of

Water,	77.17
Oil,	3.90
Dry substance,	19.93

The dry substance is composed of

Lime,	8.670
Magnesia,	670
Potash,	1.565
Soda,	1.019
Phosphoric acid,	7.784
Chlorine,	678
Silica,	1.333
Organic matter,	78.301
	100.000
Ammonia,	9.282

The fish were taken in the fall at the season when they are fat. At this season they weigh nearly a pound. Substances which abound in oil always make powerful fertilizers. The

* Third Annual Report for 1856, of the Geol. Survey of New Jersey, p. 63.

cotton seed is a well known substance, whose reputation as a fertilizer is based in part upon its oil. But fish are rich in oil, phosphoric acid and ammonia, and hence they form a concentrated manure. If the analysis is compared with those which have been given in the foregoing pages, it will be seen that the constituents of fish are admirably adapted to the purposes for which they have been employed.

§ 96. The same remark, however, applies equally well to all animal matters—flesh, bone, the hoofs, horns and hair, all are active fertilizers, their speedy influence being dependent upon the state and condition in which they are applied. Bone ground finely is much more active than when it is coarse. To obtain speedy action it must be soluble. But fish manure occupies an intermediate position—it is more speedy in its action than bone dust, but it is more transient in its effects, in which case, it has a close resemblance to guano.

§ 97. Crabs and fish of the same class have also been prepared for a like purpose. The king crab resorts at seasons of the year to parts of our coast in immense numbers. These on being taken are dried and ground when it is prepared for use. It has been sold under the name of *Cancerine* from cancer, a crab. When compared with guano, it is found quite similar in composition. As guano is supposed to owe its value mainly to its ammonia and phosphate of lime, it may be compared with fish or cancerine to determine their relative values.*

Thus Peruvian Guano contains of

Ammonia,	15.00
Phos. acid,	14.75
Cancerine ammonia,	10.75
Dry fish do	9.27
Phosphoric acid,	7.78
Phosphoric acid in cancerine,	4.05

An immense amount of fertilizing matter is lost which might be saved in the offals of fish. If they were dried or preserved

* Geol. Survey of New Jersey, p. 61, for 1856.

in a mode which should free them from offensive odor, they would be equally valuable for a manure. All the large establishments upon the extended coast of this State and upon its bays and rivers, would furnish as much fertilizing matter as is now imported into the State in guano—the cost of which is paid to foreign merchants.

At the present time, the inducements for the preservation of the offal of fish, and the taking of those fish which are not used as food are very great, in consequence of the diminished cost of transportation by railroad and the increased demand in the interior for fertilizers. The prepared cancerine for market, and which is mixed with charcoal and plaster for the purpose of removing its unpleasant odor, is composed of:*

Ammonia,	25.57
Organic matter,	29.23
Phosphate of lime,	5.90
Sulphate of lime,	10.32
Silex,	1.20
Water,	26.10
	98.32 Booth.

The king crab is used without preparation in New Jersey by the farmers of Cape May, though many are in the habit of composting them with earth. It is thus prepared as a manure for wheat, and it is stated by Prof. Cook, with the happiest effects; the poorest soils on being dressed with from two to four thousand produce from twenty to twenty-five bushels to the acre, and thirty bushels is not an uncommon crop. As this kind of manure contains but little inorganic matter, an improvement of it may be effected by the addition of ashes or lime to the compost or dirt heap. Such an addition would fit it for corn, clover or grass.

It is very possible the king crab, and fish only fit for manures, are not to be obtained in sufficient quantities upon the coast of North Carolina, to give the business an importance

* Second Annual Report of the Geol. Survey of the State of New Jersey, p. 98.

in a commercial point of view. But the real advantages of their employment is still very great, for the profits of fishing there may be added those of agriculture, which is probably neglected on account of the natural sterility of the lands upon the sounds and rivers. In many places vegetable matter may be obtained with which to form in part the compost heap, a substance which is well adapted to preserve the ammonia and other vegetable matters.

§ 98. *Concluding remarks upon fertilizers.* Husbandry in none of its branches can be conducted successfully in the absence of fertilizers. This remark is applicable only to those soils which have been under cultivation long enough to exhibit indications of incipient exhaustion. There can be no question respecting the necessity of supplying the waste of soils consequent upon cultivation, and there is no branch of agriculture which does not demand a constant supply of manures; and hence the great importance of creating enough from the immediate premises of the establishment. While it is better to purchase fertilizers than to proceed in the cultivation of the great staples without them, yet when the expenditure has to be made in cash, it is better to make composts, save the excrements of animals, under cover, procure leaves and all kind of offal, which being placed in a condition where their volatile matters may be absorbed, than to expend ready cash for those which, in the end, are no better than those made at home. To obtain the basis for the construction of compost heaps, the mud, and swamp bottoms, salt marsh-mud, when it has had time for discharging its saline matter, the dirt under buildings, which is always rich in nitrogenous matters, and many other sources may be found and used. In the eastern counties, those places in particular, which lie upon the sounds and rivers where fishing establishments are accessible, must furnish an important source of manures. The offals of fish should be composted with dirt, leaves, plaster, or fine charcoal, to deprive it of its odor and retain the ammonia. But one of the most valuable resources will be found in the decaying wood of forests, swamps and bottoms, which should be burned when there is no wind, and the ash secured under cover before it has lost a part of its potash by rains. In this latitude it is doubly necessary that all fertilizers which abound in volatile substances should be secured from the direct heat or rays of the sun, for observation very clearly proves that a great loss is sustained in all animal fertilizers, where they lie unprotected upon the ground, and especially if exposed to its direct rays. To increase the quantity of fertilizing matter upon a plantation, should be regarded as a business, and that business should be systematized. It should be followed up with the same regularity and attention as that which is bestowed upon th raising of cotton or corn. A rich plantation is agreeable to the eye; it will not wash nor become chanelled into unseemly gullies, unless the owner ploughs his grounds carelessly, or neglects to supply the immediate wants of the crop under cultivation. Exposed soils gully. Hence the importance of providing for the growth of the crop to save the soil from washing by furnishing it a sufficient protection in the crop under cultivation. There are, therefore, two considerations, either of which is sufficient to induce the planter to provide fertilizers, viz: a remunerating crop and a tillable surface, or one free from gullies. A soil as soon as it is approaching to an exhausted state, will begin to be marred and cut by streams which cross it, and those which are formed by rain. The better part is thereby carried away and lost. The tendency is to reduce the value of the plantation and render its cultivation more difficult and expensive.

The cure for all these incidental as well as direct evils, is to provide an ample supply of fertilizers.

CHAPTER XI.

Clay.—Characteristics of a good clay.—Composition of fine clays.—Composition of a clay upon Bogue Sound.

§ 99. Clay, though rarely, if ever, a constituent part of a vegetable, is still an important substance in matters pertaining to agriculture. It is one of the most important substances in construction. It is also employed largely in the manufacture of articles indispensable in the economy of the household, and is the principle material employed in the draining tile.

Clays differ widely from each other; some are fusible; others are very refractory in the fire, or scarcely fusible by the highest heat of a furnace. For certain purposes, the refractory clays are indispensable. For lining stoves and furnaces, this property should exist in an eminent degree. For household utensils, it is not necessary the clay should be highly refractory in the fire. As different properties are required for the different uses to which clay is to be put, it is desirable that the adaptedness of clay for a special purpose should be determined by methods which are within the reach of every intelligent individual; at least that good clay may be determined by some simple and easy experiment.

In the first place, good clay is homogeneous; it is free from lumps, stones and other foreign matter. In the second place, it should have an unctuous feel; this property implies tenacity, and an ability to mould readily and retain forms and shapes which is given to it by working.

In the third place it should contain sand. Too much sand destroys cohesion, but a certain proportion of sand imparts to clay an ability to dry or season. Bricks, tiles and all utensils must dry through before they can be burned, else they will crack when exposed to the heat of the kiln. Excess of sand renders moulded clay weak and unfit for handling; its tenacity will be so far diminished that it cannot be carried from place to place.

Certain clays contain so little sand that in order to dry or season well, it must be added; but when clay is to be worked by a machine, less sand is required than when it is worked by hand.

Clay that cuts smooth is probably a good clay. The surface exposed by cutting should not exhibit ragged lines, or show particles of coarse sand or hard spots.

Good clay has a uniform color, and is not spotted with ochrey matter. A clay may be red, blue, brownish or purplish, and yet possess excellent properties.

Clays for certain purposes should not effervesce with acids; this phenomenon denotes the presence of carbonate of lime, which imparts fusibility to the compound. This tendency to fuse in the kiln is increased when iron is present. All such clays will require very great care in burning, and when burnt into brick, are unfit for places where they will be exposed to great heat. Fire clays consist of alumina and a fine or impalpable sand. For withstanding high heat, as much sand must be mixed as the clay can bear and handled without breaking. Sand increases the infusibility of the mass.

§ 100. A bed of fine clay overlies the shell marl. At certain places it is fine, plastic, cuts evenly, and may be moulded readily into the form of any article in common use. On Bogue sound, it is purplish and extremely fine, and is an excellent potter's clay.

The composition of the infusible clays of the best kinds have been determined by many analyses. Thus, the celebrated Stourbridge clay consists, according to the late Prof. Johnston, of

Alumina,	38.8
Silex,	46.1
Water,	15.1
	100.0

The Woodbridge fire clay of New Jersey, according to Prof. Cook, is composed of

Water,	14.640
Alumina,	52.850
Protox iron and magnesia,	0.944
Silica,	39.76
Lime,	0.398
Magnesia,	0.650

It is one of the best fire clays in this country.

The fusible clays contain lime, iron, potash and soda, all of which vary more or less in the proportions they bear to the alumine.

The bed of clay which has been refered to, as forming one of the strata in the series of coast deposites, appears to exist in an uncommon state of purity upon Bogue sound. It is readily moulded and forms a very firm mass on drying; its grain and texture is very fine and is free from irregular lumps or regular concretions. It is, therefore, homogeneous, and is well adapted for fire-brick, tiles, etc.—and may also be employed for door knobs. It is composed of

Water,	5.70
Silex,	67.40
Protoxide of iron,	3.70
Alumina,	23.08
Lime,	0.11
Magnesia,	0.8
Potash,	0.4
Soda,	0.5

This clay contains but a small percentage of water after being exposed to the atmosphere for several months. It becomes nearly as firm as a rock. This bed of clay extends over a wide territory, and at many other points I have observed that it is equally fine and compact. It is one of the most persistent beds in the tertiary series. A fine variety of it occurs near Halifax.

Clay is sometimes employed as a fertilizer; those only, however, which are rich in lime or potash can be regarded as of sufficient importance to warrant the expense of hauling. Clays of a composition similar to the foregoing are not adapted to this purpose.

The late Prof. Johnston, in summing up the qualities of the

best tile clays, remarks that the adhesiveness of clay depends mainly upon the proportion of alumina. Clays of an average goodness will contain about 85 per cent of silex and alumina when taken together. Much depends evidently upon the coarseness of the sand, for when the sand is coarse the tenacity of the clay is very much diminished. Clays again in which the infusible ingredients is greatest, other properties being equal and favorable, are best adapted to the manufacture of good tile, besides in this case they admit of being moulded lighter and thinner. If lime and oxide of iron exist in large proportions, the clay is rendered more fusible, but in that case, it possesses an advantage of being burnt with less fuel. So with brick. The clay of the tertiary beds, it will be perceived, contains but a small proportion of lime and iron, or other elements which are calculated to confer fusibility. Hence it will probably be found that this clay will rank with the most infusible of the clays, except the porcelain clays, and being extremely fine and tenacious is well adapted to the manufacture of many fine earthern wares which are so necessary in house keeping.

CHAPTER XII.

The grasses and their functions—Different objects attained by their cultivation—Chemical constitution of the grasses—Elementary organs, and parts of the blossom.

§ 101. The grasses serve many important purposes. They clothe the earth in green, a color easy and agreeable to the eye. They protect the loose earth and prevent its washing away and transportation into the streams, or being cut into gullies. They furnish food to the beasts and birds, and the most important, the cereals, sustain the millions of the human race which now people the earth. The seed of all grasses are nutritious; the smallest are only fit for the sustenance of birds and insects. Those which are denominated corn, are those which are specially cultivated for their albuminous matters for the use of man. The latter, I do not propose to speak of under this head; the former, or the grasses, which cover the earth with green, and whose herbage forms the nutriment of cattle, compose the family upon which I propose to treat.

The diversity in kind is worthy of notice. Each one has its place. The meadow has its special occupants which usually belong to the noble kinds. The marsh and bog are covered with those which are coarse and unnutritious; and the dry hill-side, with the tough and wiry ones which serve merely the protection of the surface. The hill-side, however, has a better class of occupants; and where the surface is moist the most nutritious grow luxuriantly, and supply the herds and flocks with the most nutritious food.

It is in the temperate latitudes that the best grasses find their home, and the husbandman the best reward in their cultivation. It is in the region of the best grasses that man obtains the richest food; milk, butter, cheese, beef, pork and mutton are supplied at the least expense, where these are the material productions of the soil. Life is sustained at the least expense where the better grasses grow spontaneously. Some of them, however, must be sown and cultivated, and like the cereals be raised by the skill of the farmer. The poorest grasses frequently crowd out the better. Lands which become poor, support only the poorer kinds, and if the farmer seeks his best interest, he will displace the latter by good tillage and the use of fertilizers.

The direct objects which are sought to be obtained by the cultivation of grasses, are the production of beef, milk and butter; a greater variety of food, better in kind, and more abundant in quantity.

The indirect benefits of the grasses, in addition to the supply of food for cattle, are for furnishing a source of fertilizers for the cereals, and preserving the soil in a good condition. If cattle are left to roam at will through the ranges of forest and wild pasture, the latter object is sacrificed, though it may appear that less work or labor is consumed; still, in the long run, where lands are sold by measure, and their limits restricted by lines and corners, the losses directly and indirectly sustained more than counterbalance the gains accruing from the use of indefinite, uncertain ranges.

Another consideration bearing upon the cultivation of grasses, may be regarded somewhat in the light of a duty. Stock require a variety of food. The benefits of variety are numerous. Health is one. The appetite is cloyed by confinement. Human experience is a sure criterion by which to determine the wants of the beast. Bacon is excellent food. But who is not better satisfied with his diet, if a beef steak and a fowl help make up the routine of meals during the week? Watch the feeding of a herd of cattle or a flock of sheep, and it will at once satisfy the close observer, that they seek variety, and doing so they but follow the promptings of instinct. Grasses differ in value; while the majority of them are of the greatest importance to animals, some rank much higher in the nutritive scale than others. The most nutritive grow upon the best soils, the least either upon wet, cold soils, or upon worn out ones. Let an intelligent planter see the grass of a field, and he will tell you whether the soil is rich or poor, cold or wet. They stand as indices of thrift or poverty, industry or laziness, intelligence or ignorance.

§ 102. In the cultivation of grasses different objects are had in view. Most grasses are particularly desired for their nutritive properties, but some fulfil other functions. They may be demanded for their ability to grow in sand, when they perform the important office of confining it in its place. Some make a good turf, and their strong matted roots protect the soil and clothe the suface in a carpet of green.

That the earth may be covered, and the marshes and swamps productive in something useful to the lower forms, there are coarser grasses created which are specially fitted for such places. The Pheleum pratense, Poa trivialis, and indeed most of the rich and nutritive ones are constitutionally unfitted for the marsh. A rich, sweet grass with nutritive seeds, the

Glyceria fluctans, flourishes in the sluggish waters of streams; and what is singular, the carnivorous trout feed and fatten upon them. The broom grass, worthless as it is for stock, clothes the worn out soil and protects it from washing. It is better it should be covered even with broom grass, than burn in the sun and be washed away by the showers. Like these, all great classes or divisions of natural productions, the different families and groups have special duties assigned to them, which they assiduously fulfil, whether it be a higher and more honorable function, that of supplying nutritive food for cattle, or the lower and humbler ones, to protect a barren soil. The first perform a double office, as they protect equally well the soil beneath them; the latter is simply protective or passive. As grasses have their preferences for certain soils, as the wet, or dry, or one moderately wet, so they also require a particular climate. The Timothy grows but indifferently in North-Carolina. It requires a cooler temperature, or a less scorching sun. Upon the mountains constituting the Blue Ridge, and the adjacent ranges, it grows as well as in New England, where it is the most important of the grasses, and a source of wealth to the inhabitants. The north may have a few species which are restricted by climate; the south also has a climate which is suited to many which find the north incongenial to their constitutions. But most species of grass have wide ranges; they are less restricted when they are considered only as to ability to live, but do not grow freely; they appear under restraint and fail to make themselves of much importance.

A moist atmosphere favors development, and the production of a juicy tissue. A dry and cool atmosphere favors a dense, dry and wiry tissue, a hard outside, and a tendency to form woody fibre. Animals avoid the latter and seek the former. They are not only sweeter and more palatable, but require less effort to masticate, and less wear of the teeth, in consequence of the smaller quantity of silex in the dermal tissue.

The great variety in the constitution of grasses secures a succession of kinds for the seasons. The early spring has its kind, and a succession follows till late in autumn. Some are found fitted for food just as the snows are about to cover the ground. The farmer will not fail to profit by this succession. The early and late pasturage shortens a winter two weeks or more. The end is attained by mixing the seed of the plants we wish to cultivate. The advantage is not confined wholly to a successive supply of food, but a greater quantity grows upon a given area than if it was cultivated with one.

§ 103. The grasses proper consist of many genera, containing each many kinds or species. They constitute a very natural family of plants resembling each other in their external characteristics, and also in their internal organization and chemical constitution.

I have had occasion to speak of the chemical constitution of plants, and have called some, as the clovers, *lime plants*, and others, *potash plants*. The grasses differ from these; instead of lime or potash, they contain silica, though potash is sometimes present in large proportions, and must necessarily be present to a certain extent in combination with silica, for no doubt it is required to give it solubility.

The design and construction of the grass plant, as it was to be deficient in woody fibre, required some hard substance to sustain its slender and delicate frame. This frame work is, in a portion of the family, a hollow cylinder, or several hollow cyinders connected by impervious solid joints, sometimes called *nodes*. Others are provided with a pith as the corn stalk. Their leaves are always formed upon one plan, being long and tapering, or lanceolate with ribs running parallel with each other their entire length and never anastomosing. The middle one is stronger than the rest, and more prominent. The leaf terminates in a sheath below, which grasps or encloses the stem. The root is usually fibrous, sometimes bulbous, and creeping; it frequently becomes troublesome to extirpate as it emits roots from the numerous joints with which it is provided. The flowers or blossoms are small and never showy. They are simple, having envelopes which are in keeping with the family characteristics. Thus, there are provided two grassy outside leaves, answering to the calyx of other plants, called *glumes*, and two more delicate inner ones, answering to the coral, called *paleae*. In the centre stands the *germ*, surmounted by two feathery sessile *anthers;* and beneath and around the germ, there issues two or three filaments, or threads bearing anthers, which are little boxes containing the fertilizing matter, called *pollen*. The indian corn and several other kinds of grasses deviate from this arrangement in having the filaments, bearing the pollen boxes in a distant part, as the tassels; while the pollen receiving organs, the silks, or *pistils* are connected with the germs lower down upon the stalk. Wheat, rye, and oats, or the hollow stemed grasses, have all the floral organs in a single blossom together.

The floral organs are borne sometimes upon a *spike*, a good example of which is furnished in the Timothy grass, or wheat head, or upon a panicle, as in the oat, red top, bent grass, &c.

The grasses contain nutriment in their stalks, roots, leaves and seeds. The important part considered as food for beast, is the herbage, the stem with its leaves and head, or panicle of flowers. The seed, except in the class, cereals, is not relied upon as an article of diet. The nutiment, so called, is divided into two kinds: 1, that which contributes to the formation of flesh and muscle. 2, that which supplies heat to the system, and which is capable of accumulating in different parts of the body in the form of fat. It is designed to be burned in respiration by combining with oxygen, while the flesh producing matters supply and renew the wasting fibre.

§ 104. The value of grasses for feeding stock depends upon the quantity of flesh-forming and .heat-generating bodies which they contain. The first are known under the names of albuminous substance; albumen, the white of an egg, represents the first, and sugar or starch the second. These two classes are totally unlike each other, and cannot be converted one into the other by any known process. All substances which are used for food contain both classes, but in different proportions. Flesh of animals is the extreme of one class and fat the extreme of another. In the potatoe there is a large quantity of heat-generating matter, and a small quantity only of flesh-forming. Milk contains these two classes probably in the best proportions for young and growing animals. The cheesey matter or curd is the flesh-forming and the butter or oil the heat-generating.

In all cases it is worthy of note, that water is a very large constituent of bodies which are nutrient, even in lean meat the highest form of flesh-forming matter, about four-fifths is water.

In vegetables, especially the seed, these two classes are concentrated more than in the leaf or stem. The same bodies exist in the stem and leaves, but in less proportion. The constitution and structure of domesticated animals undoubtedly require that the flesh-forming and heat-generating bodies should be so combined and diluted with neutral ones, that in order to satisfy the appetite and fulfil the designs of nature, they should take in a bulky aliment. Hence the adaptation of grasses and herbs to satisfy the requirements of their systems. The nutritive and heat-generating substances do not differ in kind from those of the seed or even from flesh. One of the questions to be determined then, with respect to grasses, is the proportions in which these important bodies exist in them. This question is easily settled by an analysis of the plant. The starch, gum, sugar and fat represent the heat-sustaining bodies, the albumen the flesh-forming. A grass will be valuable, all things being equal, in proportion to the latter substance, or any substance which performs a similar office. Grasses which are composed mainly of silica, as the broom-sedge, are never nutritious. Those, however, which are rich in potash and the phosphates of the alkalies, are nutritious, and rank high as flesh-forming grasses. As grasses differ among themselves in these particulars, so they differ in their constituents at their different stages of growth. The stem particularly loses its nutritive properties as the seed begins to form. At this stage its woody fibre is more dense, it is less palatable, and indeed is passed over entirely by stock, and the softer vegetables consumed in its stead. Hence it is necessary in forming pasturages, to provide a variety of grasses which ripen their seed at different times, and thus furnish a juicy food during the time of pasturage. For hay, a similar rule should be observed, to supply hay which has been cut before its stalk has become woody and unnutritious. Hence, too, a meadow which is designed for a permanent mowing lot should be sown with grasses which reach the proper stage nearly at the same time. It has been common to sow Timothy and red clover together. They are, however, incompatible with each other, as the clover comes to maturity before or in advance of the Timothy. Some grass then, as a general rule, should take the place of Timothy, where it is wished to sow clover.

§ 105. Grasses grow singly or in clusters and tussocks; both frequently increase at bottom, or spread out so as to form a turf, a matting or net work of roots woven together so as to form a coherent mass, somewhat analogous to the epidermis; it is a protecting surface, spread over the loose soil so as to confine it and prevent its washing away. If grasses are mown frequently, they are more tender and soft, and under a moist sky assume the delicacy of a green velvety lawn. The grassy surface exerts an important influence over temperature, maintaining it more uniformly than if it were earthy. It prevents wide fluctuations which take place when the surface is sand, which becomes hot and burning during the day, but cold and uncomfortable during the night. The stability of the earth's surface is maintained by the grasses.

If, then, we take a proper view of the offices which the grasses perform for us and the earth, we shall set a high value upon them. We generally think of them simply as food for cattle, and it is true that in this light alone, they are of the utmost importance. But this is not all; indeed it is but a small item in consideration of the good they do and the services they perform. Though humble in their appearance and pretensions, they serve an important office in the turf, in the temperature, and in the stability and permanence of the earth's surface. To be impressed deeply with these facts, we have only to witness the moving sands of the sea-shore and the sand-storm of the desert.

Important as I have represented them, it is probable that

other forms of food for cattle will excel them in profit as food. Roots and grain outreach them on this score for special purposes at least, though cultivated at a much greater expense than the grasses. But as nature demands variety, and as the system must have food large in bulk, the place which grasses occupy cannot be filled by the more concentrated nutrients. Disease would follow if cattle were fed exclusively upon grains.

§ 106. The valuable grasses belong to several genera, in each of which there are several species.

Although grasses form one-fifth part of the flora of a country, still the number which are cultivated or domesticated is comparatively very small,—cattle consume and fatten upon plants which are not grasses, the most important of these belong to the leguminous plants, the pea family, among which are ranked the clovers. Of these, the red and white clover are the most important. The red clover is a tender plant when young, and difficult to cultivate in a hot dry climate, as many planters have experienced in the eastern part of the State.

Grasses or Graminæ, are subdivided into two great natural orders, which are known under the names of *Cyperaceæ* and *Graminaceæ*. In the former, the flowers are monecious or perfect, consisting of imbricated solitary bracts. They comprehend the coarse swamp grasses, but few of which are esteemed for fodder or food for cattle. They are, however, eaten in the spring when young and tender. The latter, have usually perfect flowers, sometimes monoecious or polygamous. The external envelopes are called glumes as already stated.

The southern genera comprehended in the family of the true grasses, are as follows:

Zizania,	Rottboellia,
Leersia,	Cenchrus,
Oryza,	Setaria,
Mulenbergia,	Tripsacum,
Agrostis,	Zea,
Aristida,	Festuca,
Cinna,	Danthonia,

Calamagrostis,	Uralepis,
Stipa,	Bromus,
Oryzopsis,	Anthoxanthum,
Spartina,	Aira,
Manisurus,	Avena,
Paspalum,	Phalaris,
Cynodon,	Melica,
Phleum,	Uniola,
Alopecurus,	Briza,
Hordeum,	Poa,
Erianthus,	Arundinaria,
Andropogon,	Eleusine,
Oplismenus,	Dactylis,
Panicum,	Elymus,
Chloris,	Monocera.

Many of the genera in the foregoing list belong to the uncultivated or wild kinds, which, though they are eaten by stock, yet are supposed to be unworthy of an attempt to introduce them into our system of husbandry.

The following list includes the cultivated species:

Botanical names.	*Common names.*
Alopecurus pratensis,	Meadow foxtail,
Phleum pratense,	Timothy or herds grass,
Agrostis vulgaris,	Red top,
" alba,	English bent,
" stolonifera,	Fiorin,
" dispar,	Southern bent,
Dactylis glomerata,	Orchard grass,
Glyceria nervata,	
Poa pratensis,	June grass,
" compressa,	Blue grass,
" trivialis,	Rough stalked meadow grass.
" serotina,	Fowl meadow,
Festuca ovina,	Sheep fescue,
" loliacea,	Slender fescue,
Cynosurus cristatus,	Crested dog's tail,
Bromus secalinus,	Willards bromus,
Lollium perenne,	Perennial rye grass,
" italicum,	Italian rye grass,
" multiflorum,	Many flowered darnel,

Avena Sativa.	*Oat.*
Avena flavescens,	Yellow oat grass,
Zea mays,	Indian corn,
Phalaris canariensis,	Common canary grass,
Anthoxanthum odoratum,	Vernal grass,
Setaria italica,	Bengal grass,
Oryza sativa,	Rice,
Sorghum vulgare,	Indian millet,
" saccharatnm,	Chinese sugar cane,
Panicum germanicum,	Hungarian millet,
" sanguinalis.	Crab grass.

Cultivated Leguminous Plants.

Trifolium pratense,	Red clover,
" repens,	White clover,
Medicago Sativa,	Lucern,
Hedysarum onobrychis,	Sainfoin.

Grasses cultivated for confining blowing sands.

Ammophila arundinacea,	Beach grass,
Elymus arenarius,	Upright sea lyme grass.

§ 107. The foregoing list of cultivated plants have been ivided into the following natural families or TRIBES:

TRIBE I.—ORYZEAE.

Oryza sativa,
Leersia oryzoides.

TRIBE II.—PHALARIDEAE.

Zea mays,
Phalaris arundinacea,
Phalaris canariensis,
Anthoxanthum odoratum,
Alopecurus pratensis,
" geniculatus,
Phleum pratense.

TRIBE III.—PANICEAE.

Panicum germanicum,
" sanguinalis,
(Includes 38 species of panicum,)
Setaria italica.

TRIBE V. AGROSTIDEAE.

Agrostis vulgaris,
" alba,
" stolonifera,
" dispar.

TRIBE VII.—AVENACEAE.

Avena flavescens,
" sativa.

TRIBE VIII.—FESTUCINEAE.

Poa pratense,
" compressa,
" trivialis,
" serotina,
Festuca ovina,
" loliacea,
Bromus secalinus,
Elymus arenarius,
(Triticum, wheat,)
Hordeum, barley,
Lollium perenne,
" italicum,

Festuca pratensis, Lollium multiflorun,
Dactylis glomerata, Cynosurus cristatus.

GRAMINACEAE.—THE GRASSES.

TRIBE I.—ORIZEA.

Containing those grasses whose spikelets are one flowered, and whose flowers are often monoecious in branched panicles.

§ 108. Oriza Sativa is cultivated only for its grain. LEERSIA *oryzoides*, rice grass, cut grass, false rice. The rice grass grows with a procumbent stem and an erect panicle, having rough slender branches and long narrow leaves, with sheaths very scabrous. It grows from two to three feet high in wet swampy places. Its spikelets are flat, and the florets of an oval form and triandrous, imbricate. Where other grasses are scarce, this may be cultivated to advantage, as it makes a good hay, and may be cut twice or three times in a season. It flowers from October to November.

TRIBE II.—PHALARIDEAE.

The spikelets *are one flowered and perfect; if more than one flowered, polygamous or monoecious.*

ZEA *mays*.—INDIAN CORN.

Probably no plant passes into or forms so many varieties as Indian corn, or furnishes so much sustenance for man and beast. It grows within the limits of latitude 42° south and 45° north, and on plains and mountains. The varieties ripen at different times, some producing in forty days from planting. Others require six months. The common eight rowed corn cultivated in the middle and northern States, comes to maturity in about ninety days. The stalk of Indian corn, if deprived of its tassel and silk, furnishes a large amount of sugar, but it does not possess qualities so agreeable as those of the sugar cane. Its ability to adapt itself to climate is of immense importance, as this property enables it to become widely distributed over the earth's surface.

GENUS PHALARIS.

Its glumes are two, membranaceous, equal, keeled and one flowered; paleae coriaceous, shorter than the glumes and pubesent at base; flowers in compound spikes.

PHALARIS ARUNDINACEA.—REED CANARY GRASS.

It has a round stem which is smooth and erect, with five or six broad leaves of a lightish green, and rough on both sides. The central rib is prominent. It grows on wet ground, and attains a height of from two to seven feet. The ribbon grass is a variety of this species. The P. arundinacea is scarcely worth cultivating for its fodder; its yield, however, during the season is quite large, but cattle are not fond of it, even when cut early and well cured. They eat it from necessity, when nothing better is furnished them. It ranks low in the nutritive scale. Phalaris canariensis is cultivated for its seed for the Canary bird.

ANTHOXANTHUM.

Its *glumes* are from two to three flowered; *lateral florets* imperfect, with one paleae, bearded; intermediate florets perfect, shorter than the latteral ones. PALEAE OBTUSE, PANICLE CONTRACTED.

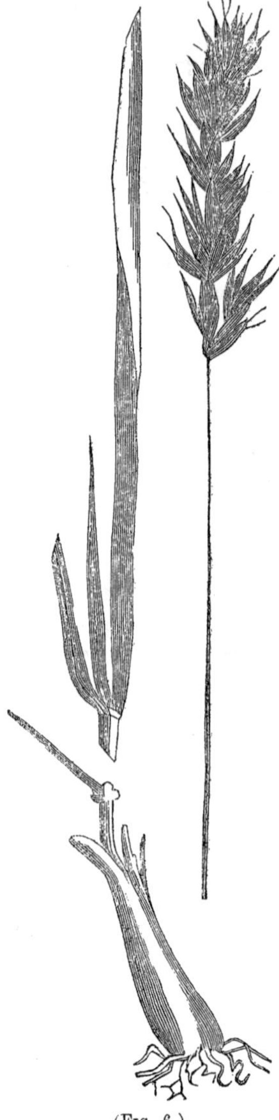
(Fig. 6.)

A. *odoratum.* *Sweet scented vernal grass.*—(fig. 6) Its stem is erect, rough at the summit, leaves hairy, sheaths striate, pubescent at the throat. Glumes are acute, hairy and membranaceous. Flowers in appressed panicles, root perennial, grows from twelve to fifteen inches high—flowers in May and June.

This grass owes all the importance which it possesses to its fragrance. It is true, that it is an early grass; and hence, may be eaten, still it is not much relished. It appears, however, that it is consumed with the other grasses among which it grows, and imparts to the milk of cows a pleasant taste, which is more particularly given to the butter.

PHLEUM PRATENSE—TIMOTHY, OR HERDS GRASS OF NEW ENGLAND—CATS-TAIL GRASS OF NEW ENGLAND.

The flowers are arranged in dense cylindrical spikes. It has two equal mucronate glumes, which are longer than the paleæ's, they include two truncate, boat shaped paleæ, without awns.

This species has an erect smooth stem, with flat linear-lanceolate leaves, whose sheaths are longer than the joints; glumes equal, ciliate and hairy root fibrous, often bulbous. Flowers in June and July, and grows best on moist lands. It grows to the height of two and a half feet. It was introduced into Maryland by Timothy Hanson, from whom it derived its name. This grass is difficult to cultivate in all that part of the Southern States which is known as the low country, or the whole of the Atlantic slope. The difficulty in its cultivation arises from the dry summers. In the months of August and September it dwindles away and finally dies out, even when protected by many large shading trees and grown upon new bottoms.

In mountainous ranges, however, it may be cultivated successfully, and as it is one of the best of grasses, it is worthy of the attention of farmers. It should also succeed in the higher grounds of the middle region.

The soil required for timothy, is one which is cool and moist, and composed of a vegetable mould, and a stiffish base of clay. On dry upland it flourishes well. On such situations it often yields two tons to the acre. It is not at all adapted to the sandy soil of the Atlantic border. The seed may be sown at two seasons: in the fall, immediately after the sowing of wheat, or in March when the ground is in an open porous state from the effects of a frost.

The quantity of seed required for an acre, is from a peck to twelve quarts. Some farmers sow only from four to six quarts. It yields in good seasons, from ten to fifteen bushels of seed to the acre, and has produced thirty, weighing 46 lbs. to the bushel, and it is worth one dollar and fifty cents per bushel. Timothy hay is preferred over all others, for horses; it is also a superior hay for working cattle in the spring.

As this species of grass gives a large product, it will be inferred at once that it exhausts the soil—especially where it is allowed to stand and ripen its seed.

The time for cutting timothy is when it has fully blossomed. At this period it possesses a larger percentage of nutriment than when its seeds are ripening. When it has stood until the seeds are ripe, the stem is hard and coarse, and is not relished so well for horses; besides, it is less nutritive, though many farmers affirm that it spends better and goes farther. Much seed may be saved from this hay, even if cut early, as all the seed does not ripen at the same time.

The old practice in the New-England States, and which is

prevalent still to a great extent, is to sow timothy and clover seed together in stocking down, after wheat or oats. This practice, however, is less common, as it is evident from the period at which the two plants ripen, that one is too immature, and if allowed to stand, the other has passed its prime. Clover is too early for timothy, and if the cutting is delayed till the timothy is ready, the clover has gone to seed, and much of its foliage has dried too much to be of any value—its stalk alone remaining green and fresh.

Wherever this grass is wished to succeed, it is highly necessary that it should not be fed too close in the fall, winter, or spring months. Hogs, if allowed to run in meadows where it is growing, will root-up and consume its bulbous, farinaceous toot, and thereby entirely destroy the crop. If cut very close to the ground, even in the northern States, it may suffer from a drought which frequently occurs about this time of the year; and a week or two of dry, hot weather succeeding immediately its removal from the field, is very liable to injure it. Although in a moist climate which prevails in mountainous regions generally, it is very easy to cultivate, yet these liabilities to fail from drouth are a drawback upon its value—though it is probably the best stock-grass which grows.

ALOPECURUS PRATENSIS—MEADOW FOXTAIL GRASS.—(Fig. 7.)

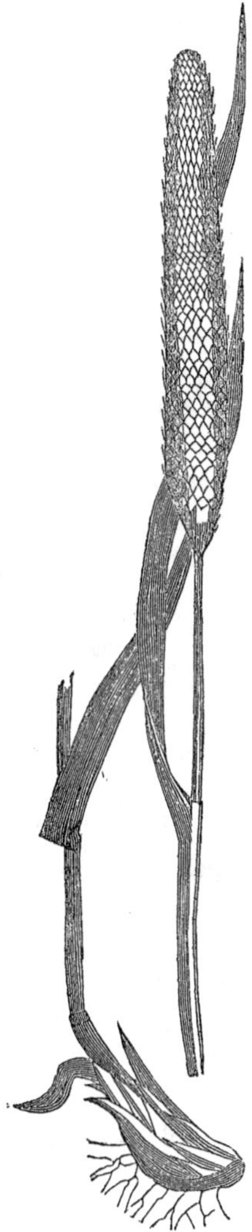
(FIG. 7.)

Its blossoms are arranged in dense cylindrical spikes, quite similar to the timothy, but may be distinguished from it by having one paleæ. Its stem is erect, smooth, and from two to three feet high. The spike is shorter than the spike of the phleum pratense, and is also softer.

This grass has received but little attention in this country. It is esteemed in England, where it is a native, though it is indigenous to nearly every country in Europe. This grass is specially adapted to pasturage, as it vegetates with great luxuriance, and starts up vigorously when eaten off by sheep or cattle. It produces seed abundantly, and hence stocks itself; moreover, it bears forcing and irrigation. It is late in arriving at maturity—requiring full three or four years to come to perfection—and hence is not well adapted to an alternate husbandry. In one or two respects it is more valuable than timothy, as it yields a large after-math, whereas the timothy yields but a small one, unless it is growing under the most favorable circumstances. Meadow foxtail forms a good sward and hence for permanent pasturage it is eminently adapted.

This grass too, is better adapted to general cultivation than the timothy as it early grows rapidly, and thrives well on all soils, except on very dry sands. It, however, thrives best on rich, moist, strong soils, and its nutritive matter increases in proportion to the strength of soil on which it is grown. It grows in the New England, the Middle States, Ohio and Maryland and it is believed that it will grow well in the Southern States, because it grows well in the warm climate of Italy. It flowers twice in the season, and the second crop exceeds the first. Sheep are fond of it, and when it is mixed with white clover, an acre it is said will yield an abundant pasturage for ten, even with their lambs. An acre, therefore, would grow grass for one cow. Loudon observes, that it affords more bulk of hay and more pasturage, than any other grass. This remark, however, may be applicable only to the climate. Another grass belonging to this genus, grows very generally in the South; it is the Floating Foxtail, Alopecurus geniculatus. Its stem is ascending, but bent at the lower joints, forming knees, smooth and glabrous; the sheaths are shorter than the joints, and it has a panicle composed of cylindrical spikes; the glumes are pubescent, but the paleæ are glabrous, with an awn at base. It grows from 12 to 18 inches high, and is common in the rice fields. It may flower as early as March. It grows in water, upon which the upper part of the stem floats. It is not so much relished by stock as to encourage its cultivation. Its early growth furnishes green and fresh food when cattle need it the most, but still it is not sought for with avidity.

TRIBE III.—PANICEAE.

§ 109. Spikelets two flowered; inferior flowers incomplete.

Panicum has two unequal glumes, the lower very small; the lower florets also, are usually staminiferous. Paleae concave, equal, beardless; seed coated with the paleae; flowers in loose scattered panicles.

PANICUM GERMANICUM.—HUNGARIAN MILLET.

The testimony which has come to hand respecting this species of millet as a fodder, is favorable, so far as southern cultivation is concerned, as it bears a drought well, and revives speedily on the occurrence of rain, and is tolerably productive on dry light soils. It becomes, however, luxuriant, only on soils which are well manured.

The plant is leafy and remains green until its seed are matured. In France its cultivation has become extended. As a green fodder, it is said to be relished by stock of all kinds.

It is sown broadcast and cultivated like other kinds of millet, and comes to maturity in about the same time. It was introduced into this country through the Patent Office.

PANICUM SANGUINALIS—COMMON CRAB GRASS.—(Fig. 8.)

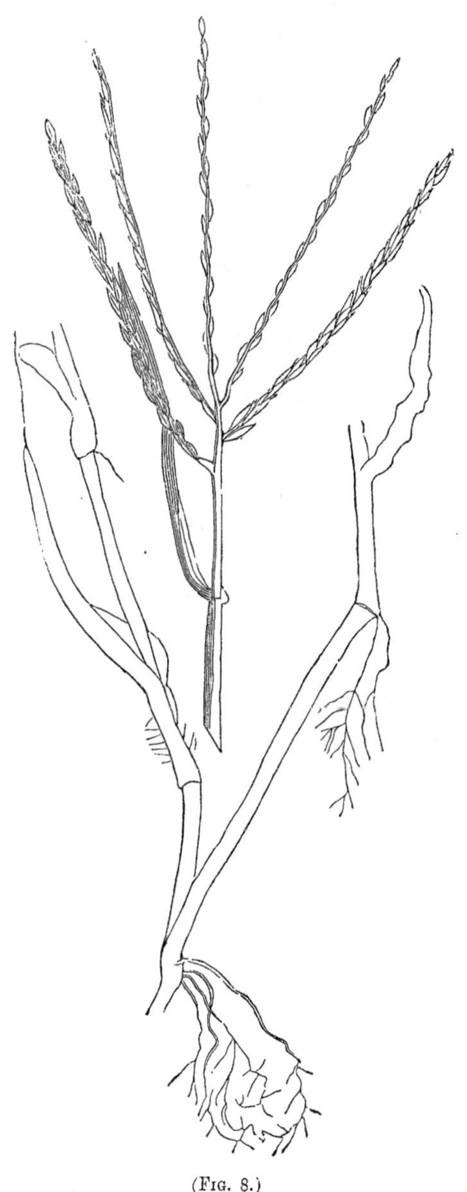

(FIG. 8.)

It has a procumbent ,assurgent, geniculate stem, which roots at the joints; the leaves are hairy, with spikes shorter than the joints. Spikes digitate, spreading, from 4 to 6. Annual, grows through the summer; common in cultivated fields. This grass, though by no means so valuable as orchard grass or redtop, still as it grows luxuriantly, and is moderately nutritious, it might justly be cultivated to a greater extent than it is at present. Cattle, horses and mules eat it with considerable relish here, and it is frequently saved for fodder. But as it is pulled up from the cornfields, it is foul with sand and dirt, and its value probably diminished. It, however, cannot take the place of the better grasses. It grows from one to two feet high in waste places, in gardens, corn-fields and yards, and is frequently a troublesome weed.

The panicum (Oplismenus) crusgalli is common about barns and waste places where the soil is rich, and some attempts have been made to cultivate it. It is rich and nutritious, and is relished tolerably well by stock, though it must be regarded as coarse fodder. There is no difficulty in cultivating this grass in this State, as it grows spontaneously in many places, and attains a height of 4 feet. It is better, and contains more nutriment than the crab grass. Its ash is composed of:

Silica,	17.325
Phosphate of iron,	0.425
Phosphate of lime,	0.625
Phosphate of magnesia,	2.831
Phosphoric acid,	6.894
Silica acid,	0.625
Carbonate of lime,	3.060
Magnesia,	2.618
Potash,	36.656
Sada,	1.885
Chloride of sodium,	5.723
Sulphuric acid,	8.524
Coal,	1.850

One hundred parts of the plant, nearly dry, gave:

Water,	4 737
Dry matter,	95.263
Ash,	11.479

Amount of inorganic elements removed in a ton of hay, 235 pounds.

TRIBE IV.—STIPACEAE.

Spikelets one flowered; inferior paleae awned ovarium stipitate. This tribe contains only wild plants.

TRIBE V.—AGROSTIDEAE.

Spikelets one flowered.

Agrostis; glume naked and beardless; two valved; one flowered; valves longer than the paleae; paleae two, membranaceous; stigmas longitudinally hispid.

AGROSTIS VULGARIS.—RED TOP—FINE TOP.—DEW-GRASS, HERDS-GRASS, OF THE SOUTHERN STATES.

Spikelets one flowered, glume naked, beardless, 2 valved, valves longer than the paleae, paleae membranaceous.

It grows erect, slender, with round smooth stems, wearing an oblong panicle; the roots are creeping. This grass, with many others of the genus agrostis, has received the name of *bent-grass*, by the English; here it is always called *herds-grass*. It is one of the most common of the field grasses, and is not so particular in its selection of the soil in which to grow, as it is found growing spontaneously in wet and dry meadows, as well as upon the dry hill side. It is regarded as possessed at least of medium qualities. There is probably no well cured hay which spends better than red top, and it is relished by stock.

The soil best suited to red top is one which is moderately moist. This grass is comparatively small, and hence does not yield so much hay to the acre, but it forms a dense bottom, and if fed close, it makes an excellent pasturage; if allowed to grow up to stalk, cattle do not crop the stems, or do not seem to relish them. Its average height is about 16 inches, but on rich soils it is twenty, and even thirty inches, and colored with a strong tinge of purple. On poor soils, it is found as low or short as six or eight inches, and is lighter colored. Some regard this dwarfed variety as distinct from the large red top of rich soils, and it frequently goes under the name of fine top.

It flowers here in June, and in Massachusetts in July. In

stocking soils after oats, or corn, the red top forms an excellent addition for mixing with clover and timothy. As the timothy diminishes the red top takes its place, and particularly does it fill the places left by the red clover as it gradually disappears.

It forms a close or dense sward, or grows thickly at bottom, and hence covers and protects the ground when the timothy fails to grow in consequence of a continued drouth. This grass should also be more extensively cultivated in this State as it is evident on examining moist meadows, it grows very well, spontaneously and without much attention

AGROSTIS ALBA—WHITE TOP.

It has an erect, round, smooth, polished stem, which is supplied with four or five leaves, whose sheaths are roughish and striate; joints numerous, from which roots are sent off when in contact with the ground. It is distinguishable from red top by its rough sheaths and the large glume toothed only at the upper part. It grows in wet places.

AGROSTIS DISPAR—SOUTHERN BENT.

The stem is large, erect and smooth, surmounted by a loose many flowered panicle, somewhat verticillate and pyramidal; exterior glume largest. It is a native of the United States. It has been commended both in England and France, but is now discarded. The hay is rather coarse, but it yields a heavy crop on good sandy bottoms which are overflowed. It tillers out and becomes strongly rooted in the soil, and hence, is a good pasture grass. It grows well in the low country of the South, where it appears to be at home.

TRIBE VI—CHLORIDEÆ.

§ 111. Spikelets in unilateral spikes from 1 to many flowered, digitate or paniculate; rachis not articulated. It contains only wild grasses.

TRIBE VII—AVENACEÆ.

Spikelets two, to many flowered, panicled; the lower paleæ bearing upon its back a bent or twisted awn.

AVENA—(OAT.)

Its glumes are from 2-7 flowered, longer than the florets; paleæ bifid, toothed with a twisted awn upon the back.

The common oat is susceptible of cultivation in high latitudes, where it is the most profitable grain. In warm climates bears a lighter grain. The stem of the oat is quite nutritious, and forms, with meal, an excellent feed for horses.

The oat plant when sun-dried,

Contains water,	9.53
Ash,	2.37
Calculated dry,	2.61

The ash of the straw, consists of

Silica,	13.399
Earthy and alkaline phosphate,	8.902
Carbonate of lime,	7.254
Magnesia,	0.448
Potash,	60.035
Soda,	3.622
Sulphuric acid,	5.754
Chlorine,	0.581

This analysis was calculated without carbonic acid or organic matter. These amounted to in carbonic acid 6.140; organic matter 2.400.

In a ton of straw there will be removed from the soil in,

Silica,	21.907 lbs.
Phosphates,	14.555
Carbonate of lime,	11.868
Magnesia,	0.732
Potash,	98.157
Soda,	5.921
Sulphuric acid,	9.408
Chlorine,	0.950
	163.498 lbs.

The amount of ash in an unripe straw is greater than after it has ripened, which is undoubtedly owing to the transfer of matter from it to the grain. The ash of an unripe straw amounted to 3.15, which calculated from a perfectly dry straw, amounts to 3.48.

The oat is an exhausting crop to soil, but for that reason it should be widely cultivated where the climate suits it. It is for this reason that it is so valuable for foód, both for man and beast.

In this family we find the

AVENA (DANTHONIA) SPICATA.—WILD OAT GRASS.—(fig. 8.)

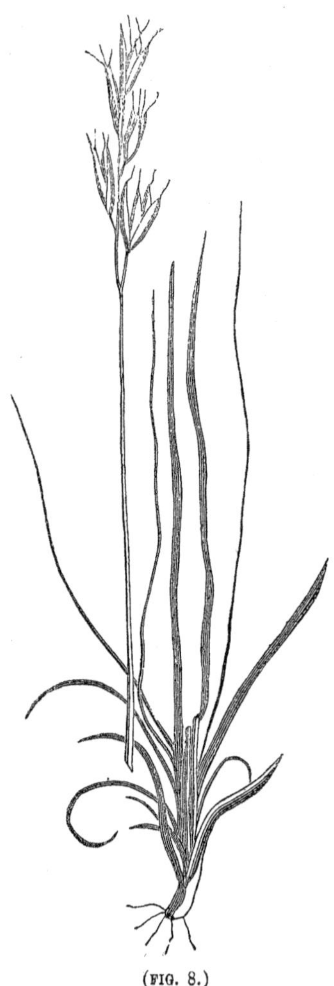

(FIG. 8.)

It has an erect pubescent stem, and tubular pubescent leaves, with sheaths bearded at the throat. Glume usually six flowered, longer than the spike margins membranaceous. Paleae two; exterior one lanceolate villous, the sides terminating in two awns, with the spiral one upon the back. Common in the middle country from Carolina to Georgia.

It grows in dry sunny pastures, aud attains a height of twelve to eighteen inches. It is of but little value for pasturage or hay.

AIRA FLEXUOSA—WOOD HAIR-GRASS—COMMON HAIR-GRASS.—(fig. 9.)

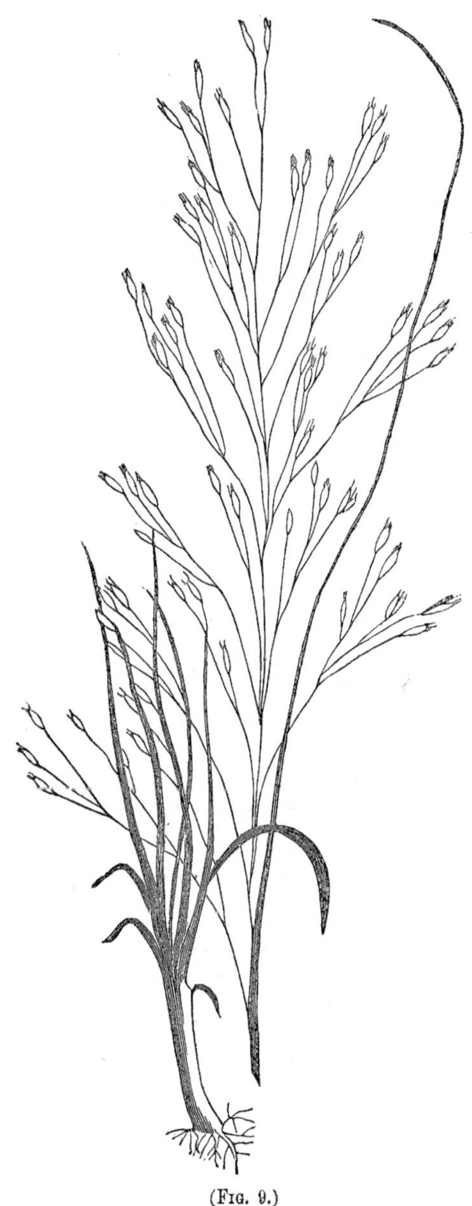

(FIG. 9.)

It has an erect, terete, glabrous stem, with setaceous leaves and a diffuse panicle, whose branches are somewhat verticillate; glumes unequal; paleæ equal, exterior one pubescent at base, and bearing also an awn. The grain is oblong and smooth. It flowers in August and September. Figure taken from the grass when in fruit. In high dry pastures, it grows remarkably well, and is eaten freely by sheep. It is poor in nitrogen, and is worth nothing for cultivation.

TRIBE VIII.—FESTUCINEAE.

Spikelets two to many flowered; panicles sometimes racemose, and generally without awns.

POA.—(MEADOW GRASS.)

The poas have two glumes, and usually many flowered. Spikelets compressed; paleae sometimes woolly at base; scales smooth; panicle more or less branching or scattered.

POA COMPRESSA—BLUE-GRASS—WIRE-GRASS.

Stem decumbent and compressed, ascending and surmounted with a dense compressed panicle, somewhat onesided, and provided with short bluish green linear leaves. Spikelets flat ovate oblong, and from four to nine flowered, which are rather obtuse, and hairy below the keel. It rarely exceeds 14 inches in height. It has a creeping root and a geniculate stem, and much compressed, and under favorable conditions grows to the height of 17 or 18 inches.

The blue grass varies much in its appearance. On dry soils it grows in tufts with rigid culmlike or wiry stems; it is also short, and has small compressed panicles, and the whole plant has a bluish green color. It is solid and heavy, and also tenaceous of life as might be suspected from its growth upon very dry knowles, and in wheat fields is frequently regarded as a pest. It is, however, a very nutritious grass, and is eaten freely by stock. It is valuable as a pasture grass.

POA PRATENSIS—SPEAR-GRASS—GREEN MEADOW-GRASS—JUNE-GRASS—KENTUCKY BLUE-GRASS.—(Fig. 11.)

(Fig. 11.)

Stem smooth, erect, terete, surmounted by a rather spreading crowded panicle, and whose spikelets are ovate, acute and crowded on the branchlets, from two to five flowered. Glumes unequal, sharply accuminate, lower paleæ five nerved.

This grass is a native of Europe, but has become extensively naturalized in the United States, both north and south. It is particularly at home in some of the south-western States, as Kentucky and Tennessee. It extends through the Atlantic States as far south as Charleston, where, according to Elliott, it grows to the height of 18 inches, where it also makes a fine winter grass, remarkable for its deep green color, and soft succulent leaves. It bears the summer heats in close, rich soils, and wants only size to render it one of the most valuable acquisitions to the farmer. It is perennial, and hence deserves the special attention of the southern planter, as there is a great want of good perennial pasture grass. Nor is there the least doubt but that it can be generally cultivated in the eastern and midland counties of the State. As for the western counties, no farther proof is required than what is already known of its ability to thrive there. This grass continues green and fresh in Western New York, frequently as late as December, it is probable, therefore, that in a large portion of Western Carolina, it will continue growing most of the winter. Although it continues to grow during a long period, yet it sends up its spike of flowers but once in the year, which, in this climate is from about the first of June to July. It continues afterward to spread at the bottom and furnish a thick mat or growth of leaves. It, therefore, makes a good turf. It is not so particular in its selection of soils as it grows on dry knowles as well as moist places. But still it flourished best in a good soil, but here it is important to obtain a grass which will endure a drought and grow on poorish soils.

The produce is ordinarily small, but it is of a fine quality. For lawns and door yards, it is probably better adapted than any grass in cultivation. One of the difficulties it has to contend with in this State is its consumption by the hog. This would not be so formidable to surmount if it attained perfection at an earlier period, requiring two or three years to get perfectly set.

As it requires time to attain perfection, it is not well adapted to an alternate system of husbandry, or when land is to be ploughed every two or three years. Shaded pastures furnish the best examples of this grass in Kentucky where it ripens its seed about the tenth of June. In August it takes another vigorous shoot and continues to grow till stopped by the cold of winter. When it dries up in the drought of summer, it is still nutritious. It continues to furnish under the snow pasturage for mules, horses and sheep.

If designed for hay, it should be cut late in flower, and if mixed with clover, the yield will be at least midling in quantity. It is eaten and relished by all kinds of stock. It seems, however, to flourish best on what are called limestone soils, similar to those of the Kentucky limestone belt. It is maintained by several writers that the June grass is deficient in nutritive properties, that it is far inferior to timothy; yet cattle do fatten upon it, and so far as observation goes, the cattle that are raised and prepared for market in Kentucky, are equal to any grass-fed animals seen in market. Prof. Way, whose analysis of this grass, have led to the unfavorable opinions respecting its deficiency in flesh-forming elements, may have analyzed specimens, which, growing in England, may not have been as nutritive as those commonly growing in our climate. It is certain that the composition of plants are very variable under different circumstances, soils. etc.; variable also at the different periods of growth.

In Kentucky farmers sow in September or February. Some prefer a late winter or early spring sowing to save the tender plant from frost. It is sown both in open ground and woodland. If sown in woodland it should not be grazed until it matures seed. The seed is often mixed with timothy and clover, and half a bushel of the seed of June grass is sufficient for an acre. By mixing, the field may be fed at an earlier day. Ultimately, the June grass takes full possession of the field.

POA TRIVIALIS.—ROUGH MEADOW GRASS.

Stem or culm somewhat scabrous; leaves smooth; narrow with scabrous sheaths; panicle equal and diffuse, somewhat verticillate. Spikelets three to four flowered; glumes unequal; scabrous at the apex; lower paleae obtuse; pubescent at base; culm from two to three feet high.

In England this grass is highly esteemed, and according to the opinion of Mr. Curtis, an English writer, it is one of the most valuable, both for hay and pasturage. In this country, however, it does not stand so high in the estimation of agriculturists, but it is probable that it has not been so fairly tested as the blue grass. Mr. Sinclair recommends it, and says of it that it is superior in produce to many other grasses; it is nutritive, and oxen, horses and sheep exhibit a marked partiality for it. It grows vigorously only on moist situations; when upon dry pastures it is only inconsiderable in quantity.

He, (Sinclair,) remarks that it should be mixed with other grasses, when it will nearly double itself, which is in consequence of being partially sheltered. Where spots in pastures are closely eaten down it will be found the places were occupied with this grass, proving thereby the fondness of stock for it. It is not so widely diffused as the June grass, but it is found in Kentucky, from which it may be distinguished by its rough sheaths. It has a fibrous root and is an annual. It should be cut when in seed. It has more nutriment in its aftermath than when cut in seed. In a specimen which I submitted to analysis, I found:

Water,	77.374
Dry matter,	22.626
Ash,	2.078

This was cut the 8th of June, was thirty inches high, and in flowers, having attached its radical leaves.

Another species which was younger and cut May 13, just heading out, gave:

Water,	81.564
Dry matter,	18.436
Ash,	2.267

Another, at about the same stage of growth, cut May 20, gave:

Water,	80.75
Dry matter,	17.91
Ash,	1.34

The analysis, however, was confined to the stalk; the leaf of the stalk gave:

Water,	75.50
Dry matter,	21.56
Ash,	2.84

In three trials for the quantity of ash in plants growing in this country the quantity exceeds that obtained from the plant growing in the climate of England. Prof. Way obtained ash 1.95. The June grass contains, according to Prof. Way:

Alluminous or flesh forming elements,	10.35
Falty matters,	2.63
Heat producing elements, consisting of starch, sugar and gum,	43.06
Woody fibre,	38.02
Mineral matter, or ash,	5.94

The latter is calculated from the dry substance. The ash of the June grass which I submitted to analysis, gave:

Silex,	56.320
Earthy and alkaline phosphates,	14.980
Carbonate of lime,	3.540
Potash,	15.624
Soda,	6.828
Magnesia,	1.996
Sulphuric acid,	.200
Chlorine,	.863
	100.351

The plants were selected from well made hay.

POA SEROTINA—LATE FLOWERING MEADOW-GRASS—FALSE RED TOP—FOWL MEADOW.—(Fig. 12.)

(Fig. 12.)

Stem and leaves smooth. Panicle elongated diffuse, branches in fives or sixes whorled. Spikelets ovate, accuminate three to four flowered, tinged with yellow at the apex ; glumes long, lanceolate, very acute ; paleae lanceolate, rather obtuse and pubescent at base.

The leaves are 2.63 lines wide, and 4 or 5 inches long ; root, perennial. Flowers in July. Ripens about the first of August, and becomes drooping.

It grows best in moist places or meadows, and yields abundantly. Its hay is excellent; sheep and other stock eat it with avidity and thrive, especially if mixed with clover. It is highly esteemed in Europe. It grows well in the southwestern States. Some think it superior to Timothy as its culms are more tender.

It grows in all parts of New England and New York, and is esteemed by all for its qualities. It is quite productive. It grows three feet high, and is liable to lodge or fall down in consequence of its slender stalk.

There is no doubt this fine grass may be cultivated in the low rich grounds of the eastern counties, particularly in parts of Hyde county.

The genus Poa contains a large number of species which inhabit woods and woody places, or high and mountainous regions. Although known to be relished and eaten by cattle, they do not yield enough to make it an object to introduce them into the cultivated fields. Thus, the Poa nemoralis, wood meadow grass, is a good grass so far as its properties are concerned. It has been recommended for cultivation by Sinclair, who remarks that, although the produce is inconsiderable, yet its early growth in the spring, and its remarkably fine succulent herbage, recommend it for admission into company with others which form good pasture grasses. For hay it is not recommended as its yield would be too inconsiderabe to deserve attention. It flowers early in May.

POA NERVATA.—(Fig. 13.)—MEADOW SPEAR GRASS—FOWL MEADOW OF SOME FARMERS.—NERVED MANNA GRASS.

(FIG. 13.)

The stem is slightly compressed—bears an open or spreading panicle, with small ovate, oblong and green spikelets—leaves in two rows, and rough, and grows from two to three feet high.

This American grass is highly nutritive. The ripening of the seed does not diminish the nutritive value of the stem and leaves. It is hardy, grows best in most places. It is eaten by cattle both in summer and winter, but is more relished in the latter than in the former season.

FESTUCA.

Glumes two, unequal, many flowered. Paleæ two lanceolate; outer one accuminate, or awned. Panicle usually compound.

FESTUCA OVINA—SHEEP FESCUE.—(Fig. 14.)

(Fig. 14.)

Stem slender, surmounted by small panicale, with spikelets from two to six flowered; awn inconsiderable; leaves, bristle shaped, reddish or greenish. It grows from 6 to 10 inches high, in dense perennial rooted tufts. It flowers in June and July; grows in dry pastures, and makes an excellent pasturage for sheep.

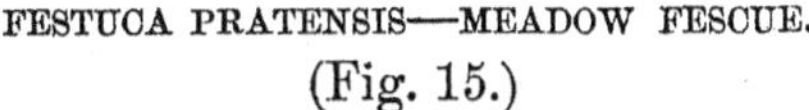

FESTUCA PRATENSIS—MEADOW FESCUE. (Fig. 15.)

Its panicle is branching, nearly erect, slightly one-sided, and with linear spikelets, and with from five to ten cylindrical flowers; color of the leaves of a glossy green, lower ones broad and pointed and with roughish edges, root creeping perennial. Flowers early in June. It grows in rather wet open grounds to the height of two or three feet.

The qualities of this grass giveit a tolerable high rank among the pasture grasses. It has long tender leaves, which are relished by cattle. It sometimes forms a good turf in old pastures. When sown, it should be mixed with orchard grass, June grass, or common spear-grass.

The figure was taken from a specimen near its maturity, and past flowering.

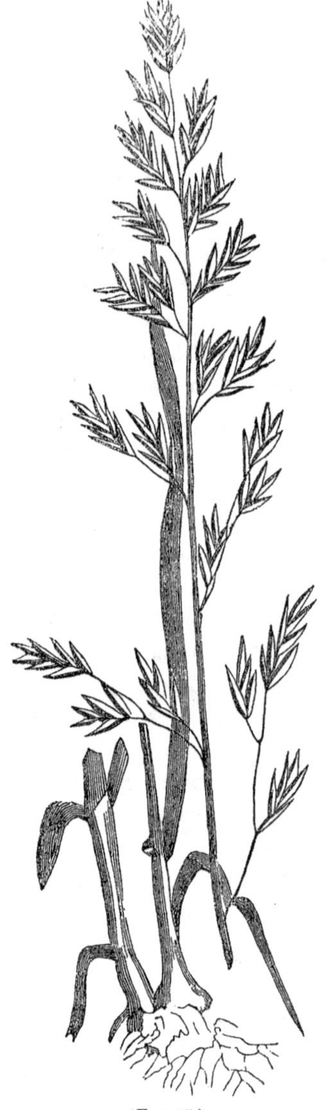
(FIG. 15.)

FESTUCA LOLIACEA—SLENDER-SPIKED FESCUE.

Stem erect, slender; spikelets acute, close pressed, rather crowded, and from ten to twelve in number. It grows in moist meadows in small tufts, root perennial. It is a nutritive grass, and would form good pastures, but it is too rare to be ranked among those worth cultivating.

The fescue grasses are common in most meadows, and occupy shady as well as sunny places; among the most valuable and common of the tribe, is the Festuca pratensis. Its stem is round and smooth, and from 2 to 3 feet high, with creeping roots, and surmounted by an erect branched panicle, and somewhat one-sided; spikelets linear, with from five to ten flowers. The leaves are long glossy green striated, and have rough edges.

Flowers in June and grows in moist pastures. It ripens its seeds early, and hence takes possession of the ground before other grasses are matured. It is a nutritive plant, growing in stiff moist soils, and in shaded places. Darby does not speak of it as a southern grass.

BROMUS.

Glumes two, many flowered, and shorter than the florets; florets imbricate in two rows; lower paleæ cordate emarginate, and sometimes armed with an awn below the summit; scales ovate smooth.

BROMUS SECALINUS—CHESSCHEAT.—(Fig. 16.)

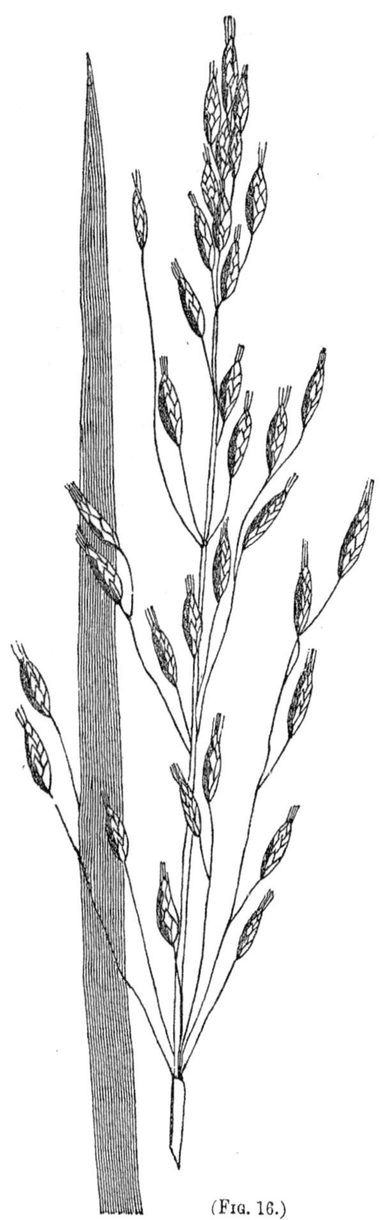

(Fig. 16.)

Stem glabrous, erect, swollen at the joints, leaves ciliate, pubescent on the upper surface. Panicle branching erect or nodding; spikelets compressed oblong ovate, florets about 10 longer than the bristles.

The remarkable views which are entertained of this plant, excuse the notice of this worthless grass in this place. It has been a common opinion with a very large proportion of farmers, that wheat changes into chess, the grass under consideration. This has frequently been, in one sense, favored by the fact that when wheat has been winter-killed, chess has sprung up in its place, therefore, to those who have not been careful observers, it has seemed that the wheat itself has undergone the change which they maintain; usually, this view seems rational, because chess has not been observed by them in this particular place in former times. Notwithstanding this apparent support to the doctrine, it only requires a good eye to detect chess in almost any corner of a cultivated field, and if it has not appeared before on a particular spot, it has probably been owing to the fact that it has been occupied by other plants and grasses which exclude it.

Facts, when properly ascertained and sifted, never sustain the doctrine of a change of one species to another. There is in nature no transmutation of the kind. Northern Indian corn after growing in the south for a few years, assumes the habits and appearance of southern corn, which is a thing quite different from the one under consideration, the change of one species into another. Chess, though it possesses some nutriment, yet it is too low to encourage its propagation. It is rather a pest which should not be allowed to mature seed, and thereby propagate itself among the valuable grains and grasses. It is an annual grass, but if cut early, will spring up and propagate itself the succeeding year.

COCKSFOOT GRASS, ORCHARD GRASS, DACTYLIS GLOMERATA.

(Fig. 17.)

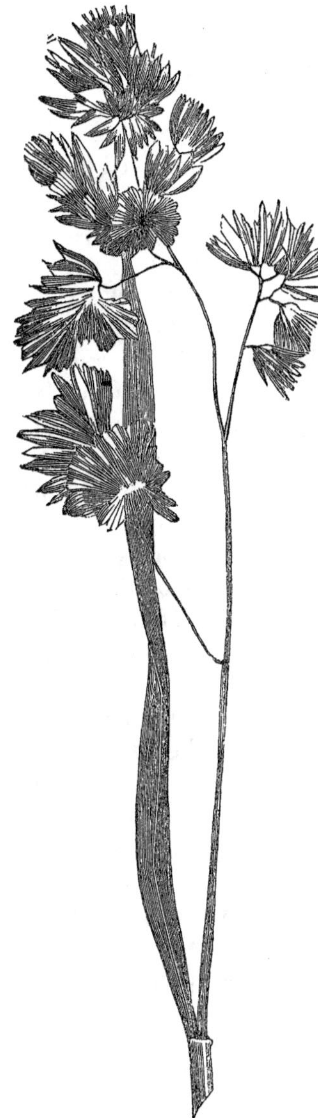

(Fig. 17.)

Flowers in dense tufts or spikelets, crowded in clusters, one-sided, with a dens branching panicle at top. It grows erect and attains a height of three feet; not perennial; it is a native of Europe, but has been naturalized in many parts of this country, and Elliott says that it has become naturalized on James Island, near Charleston, South-Carolina. This being the case, furnishes sufficient evidence that it is an important grass for the South.

The orchard grass is very widely distributed. It is well known in the north of Africa, Europe, Asia and America. It is said that it was introduced into England from Virginia, where it now forms one of the most common grasses of English pastures, is highly esteemed among cattle feeders, being exceedingly palatable to stock of all kinds.

This grass is worthy of culture from its rapid growth, luxuriant aftermath, and its endurance of close cropping, and when fed down closely it recovers in a shorter time than any other grass under cultivation. It forms an excellent grass for mixing with clover; it is free from the objection which applies to the case of timothy, as it reaches its mature state about the same time as clover. Hence, it will be perceived that it is an earlier grass. The time for cutting it for winter food is when it has blossomed. If delayed until the seeds have ripened, it is far less valuable, as it loses at this stage its juiciness. Thick tufts of it form in pasture lands, when it is not fed close. As it regards resistance of drouth, it is well known that it bears it well, in which respect it is quite unlike the timothy. Good observers declare that it produces more pasturage than any other grass. On this point the opinion of the late Judge Buel, of Albany, coincided with other eminent agriculturists, and all agree in two other important points, viz: that it should be kept fed close and that when it has had only five or six days to recover, it acquires a good bite for cattle. These points give it a preference again over timothy. Sheep are more fond of it than any other grass. It is less exhausting to the soil than many other nutritive grasses, which arises from the lightness and small amount of seed which it produces. A bushel of seed weighs only twelve or fourteen pounds. This grass is but little cultivated in New England, probably from the preference given to timothy and red top, which is rather remarkable, seeing so much hay and pasturage is required. One of the finest fields of grass the writer ever saw was upon the plantation of Col. Capron, at the Laurel. Orchard grass, when sown sparingly and upon uneven ground, is disposed to grow in tussocks. This fault may be remedied by preparing the ground properly and sowing a sufficieut quantityy of seed. This grass, however, should not be cultivated by itself, unless it is wished to grow it for seed. The celebrated Sinclair gives a formula for the formation of a crop for pasturage. He mixed the seeds of certain grasses in the following proportions:

Doctylis glomenata,	4 pecks.
Festuca pratensis,	3 do.
Timothy,	¾ do.
Fiorin, or agrostis stolonifera,	1 do.
Holcus areñaceus,	2 do.
Lolium perenne,	3 do.

Poterium songuisorba, (burnet)	2 pecks.
Trifolium pratense, red clover,	6 lbs.
" repens, white clover,	8 do.

This mixture was regarded as sufficient for an acre. We see in this prescription a love for variety and an excessive amount of seed. As pasturage is one of the great desiderata in this State, and as this grass stands dry weather remarkably well, it will probably be one of the most important measures in husbandry to encourage its cultivation. Whether it can be shown hereafter that it will give as much profit per acre as has been reported for a field near Rochester, N. Y., can only be determined by experiment. The profits reported as having been reared from one and a quarter acres of ground were given in the Genesee Farmer, Vol. V, p. 245:

There were obtained 17 bushels of seed, $2 per bushel,	$34 00
Yielding, also, 2 tons of hay, $10 per ton, for the first crop.	20 00
There were obtained 1½ tons for the second crop,	15 00
Amounting to	$69 00

Expense for gathering crops:

Cutting and shocking seed, one hand half a day,	0 50
Threshing,	1 00
Cutting stuble,	1 00
Making the same into hay and overhauling,	1 50
Cutting and making hay of the second crop,	2 00
Interest on the value of land,	4 87
	$10 87
Deducted from sales, leaves a nett gain of	58 12

To save the seed properly requires the skill of a good cradler, who cuts the tops and ties them in bundles to dry in the field for eight or ten days. They should be hauled into the barns and threshed immediately with a flail. If there is a large quantity of seed it should be still allowed to dry upon the floor, as when retaining moisture it is apt to heat in the heap, when the vitality of the seed is destroyed. The seed, as above stated, is very light. If sown with clover, one bushel of orchard grass to ten quarts of clover seed makes the proper preparation per acre. If sown alone, two bushels are required. For pasturage alone, a mixture of the white clover will form an excellent addition. Whatever opinions may prevail with respect to the cultivation of the grasses in the eastern part of the State, or even the middle, there can be but little doubt, that when the attempt is made to introduce a more extended pasturage, this grass will have the preference over many others.

The analysis of the ash of the orchard gave, Prof. Way:

Silica,	26 65
Phosphoric acid,	8.60
Sulphuric acid,	3.52
Carbonic acid,	2.09
Lime,	5.82
Magnesia,	2.22
Per oxide of iron,	0.59
Potash,	29.52
Chloride of potassium,	17.86
Chloride sodium,	3.09
Percentage of ash furnished by the dry plant,	5.51

The nutritive value of this grass is exhibited in the following analysis of Prof. Way:

Water,	70.00
Albuninous matter, (flesh forming,)	4.06
Falty matters,	0.94
Starch gum sugar, (heat producing bodies,)	13.30
Woody fibre,	10.11
Ash,	1.59

ELYMUS—WILD RYE.

It has two or more spikelets at the joints of the rachis, and is from 3 to 9 flowered. Glume 2, nearly equal, sometimes wanting; lower paleae entire with a short awn;—upper one bifid. Scales ovate hairy.

ELYMUS ARENARIUS.—UPRIGHT SEA LIME GRASS.

Stem erect, round, smooth from two to five feet high, and bearing sessile spikelets; leaves long, narrow, rolled inward, and rough on the inner surface; root, long, perennial and creeping.

Resembles beach grass in its mode of growth; it is also a valuable grass for confining blowing sands.

In England it is called the sugar cane, from the quantity of sugar in its stem.

The E. virginicus, (wild rye,) E. canadensis, (canadian lyme grass,) E. striatus (slender, hairy lyme grass,) grow along the banks of rivers and streams, but they are of no special value for cultivation.

LOLIUM.

Spikelets many flowered, solitary on each point of a continuous rachis, placed edgewise.

LOLIUM PERENNE.

Stem erect, smooth, leaves flat, acute, smooth on the outer surface, roughish on the inner, glume shorter than the spike, flowers from six to nine, awnless. Flowers early in June. From 15 to 24 inches high. Root perennial, creeping.

This is regarded as valuable grass both in England and France. It is relished by stock previously to its blossoming, afterwards it becomes hard and less palatable.

It is not equal to the orchard grass in any respect, but at the same time it must be admitted that it could not have stood its ground so long in England and France unless its merits are considerable. It is doubtful whether it can be cultivated in this State with profit. It seems to attain perfection in a more humid climate than ours.

LOLIUM ITALICUM—ITALIAN RYE-GRASS.

It is inferior to our best grass, as timothy, orchard-grass, blue-grass, etc. In some points of view, however, it is superior to them, as it may be cut several times, when sown upon moist rich land. It grows luxuriantly, and for soiling cattle it is an excellent addition to our grasses, as it bears cutting well. Its actual value to us, however, is still to be determined by farther experiments.

LOLIUM MULTIFLORUM—MANY-FLOWERED DARNEL.

This grass is so little known in this country, that it may be passed over without remark.

TRITICUM—WHEAT.

Flowers in spikes; spikelets imbricate sessile; $\frac{3}{4}$ flowered. Glume two, nearly equal opposite; paleæ lanceolate; the lower concave acaminate awned; scales two ciliate.

Wheat is supposed to have been indigenous to Central or South-western Asia. It is known to have been cultivated from the earliest times.

Like the Indian corn its varieties are numerous, amounting at the present time to about 300, which are known to be under cultivation.

The characters of these varieties are essentially the same. The modifications affecting merely its appendages without extending to its essential characteristics. The character of the soil influences the value of the grain; it is always richer and better on rich substantial soils. When grown upon those which abound in vegetable matter its grain is light.

TRITICUM REPENS—COUCH-GRASS—SWITCH-GRASS—DOG-GRASS—DUTCH-GRASS—QUACK-GRASS.

It has an erect stem, with smooth joints, two upper most remote; spikelets close pressed, leaves acute, upper one broadest; sheaths striated, roots creeping extensively. Introduced from Europe; flowers in June.

This grass is cut in blossom,—is relished by cattle, and makes a nutritious hay. In gardens and other cultivated grounds it becomes a great pest, from the difficulty of eradicating it. Its roots are short-jointed, and send out fibres from all of them, in consequence of which it grows and maintains itself when a single joint remains, besides it is tenacious of life, and does not readily die when left upon the earth's surface.

This grass cut in May 13, gave,

Water,	81.564
Dry matter,	18.436
Ash,	2.367

A second specimen from the same bed, cut, June 8, gave,

Water,	77.374
Dry matter,	22.626
Ash,	2.073

As this grass approaches maturity, its inorganic matter decreases and its woody fibre increases. A third specimen taken when in full blossom, gave,

Water,	68.50
Dry matter,	30.50
Ash,	1.00

An analysis of the ash of this grass, gave me,

Silica,	27.150
Phosphates of lime, magnesia and iron,	19.250
Lime,	0.112
Magnesia,	trace
Potash,	10.350
Soda,	26.985
Chloride of sodium,	8.990
Sulphuric acid,	4.811
Carbonic acid,	1.455

The same change takes place in the lolium perenne. These experiments have an important bearing on the time they should be cut for hay. It is well known that stock relish grass and hay while it is succulent and juicy. After the woody fibre is largely formed it is less palatable and more difficult to masticate; besides, it wears the teeth more, and less nutriment is taken into the system.

CYNOSURUS CRISTATUS—CRESTED DOG'S-TAIL.

Its stems are about one foot high, stiff and smooth, provided with fibrous perennial root, more or less tufted. Its stem being hard and wiry, cattle usually refuse to eat it. In dry sheep pastures, it is more valuable as a permanent grass. Its stem is used in the manufacture of straw plait.

The common broom-sedge is another grass whose stem and leaves become hard and wiry with age, and still more unfit for food for cattle than any of the preceding. It takes possession of old and worn out fields, and imparts to them a look of barrenness, which, in many instances, they do not deserve. Cattle eat this grass only in the spring, when it first springs up, and when it is comparatively tender. Although almost worthless for fodder when mature, it is still better for the ground to be covered and protected by this grass than to be naked and exposed to the heat of the sun and the action of rains. This grass has but a small proportion of nutrient matter; at the same time the consideration how fields should be treated when covered with it, is worth a moment's consideration. When such a field is to be ploughed for a crop of wheat, it is important to lay it under while it is still green, or before it has reached its full maturity. At this period it is more valuable as a fertilizer; the proportion of silex in the stem being relatively less and the more valuable elements are greater. When mature, it contains about 72 per cent. of silica, and only 8 per cent. of the phosphates of lime and magnesia. The only grass which approaches this in its mature state in the proportion of silica, is the Italian rye-grass, which contains 60 per cent. In burning off a crop of broom-grass, a large proportion of this silica becomes insoluble. Hence it should be ploughed under when well grown, when all its nutritive elements are in the best condition to aid the growth of the succeeding crop.

CHAPTER XIII.

Red clover belongs to the Leguminosæ—Organic constitution—Composition of its ash—Differs in composition from the grasses—Failures in its cultivation—For a green crop—Lucerne—Sanfoin.

§ 113. In the northern and western sections of the United States the red clover, though not a grass, is now regarded as one of the important resources of husbandry. It forms of itself an excellent food for cattle. It is one of the most speedy and effectual means by which soils may be brought to produce remunerating crops. It is therefore both a nutriment direct for cattle, and a fertilizer for the cereals. It is in virtue of its rapid growth, large herbage and roots that it occupies a place in husbandry so important; besides, it derives no inconsiderable part of its substance from the air. In the natural classification, it belongs to the family *leguminosæ*, or the same family as the bean and pea. Its common name, clover, is most in use. It is sometimes designated by the term *trefoil*, three leaved.

It scarcely requires a description, as it is known by every farmer and planter. Its stem is inclined to be prostrate or ascending, and the leaves are oval, and stand in threes at the termination of the stem.

The red clover, after many years cultivation, has developed a number of varieties. One of these varieties is biennial and another is perennial, and like many other biennials which has become so in other families of plants, it frequently lasts three or four years, provided it is not suffered to go to seed.

Clover is a very easy plant to cultivate in a cool, moist climate. In one similar to North Carolina, which, perhaps, is more subject to droughts than New England or New York, it is more difficult. This arises from the tenderness of the young plant. In its early stage, if exposed to a burning sun, it dies. But it is not difficult to protect beneath the shade of another plant, and thereby save it from perishing.

Clover is a nutritious fodder, and cattle and horses are very fond of it. But as it frequently grows very rank, it is not perfectly cured, and in a green state it moulds. If fed to a horse in this condition, which is at all inclined to the heaves, it will certainly produce it.

As a nutriment, clover takes rank with the best of grasses. According to Prof. Way, red clover contains,

Water,	81.01
Albumen,	4.27
Fatty matters,	69
Gum, starch, sugar, or heat-producing principles,	8.45
Woody fibre,	3.76
Ash,	1.82

Clover is a lime plant, but this element increases with its age. In the young plant the proportion is much smaller than in the old. Thus:

	OLD.	YOUNG.
Silica,	0.850	0.981
Phosphates of lime, and magnesia, etc.,	20.600	30.245
Carbonate of lime,	30.950	7.642
Magnesia,	3.930	2.285
Potash,	25.930	33 688
Soda,	14.915	7.164
Chlorine,	1.845	3.642
Sulphuric acid,	0.495	6.723
Carbonic acid,		5.744

The upper part of the stem, with the leaves and heads, gave a composition varying from the above, thus:

Silica,	0.810
Phosphates,	21.900
Carbonate of lime,	32.333
Magnesia,	0.200
Potash,	27.940
Soda,	6.753
Chlorine,	3.780
Sulphuric acid,	3.366
	98.682

From the foregoing analysis it will be perceived that clover differs in composition from the grasses. It contains only a small per centage of silica; and hence, cattle and horses masticate it easily. Two elements exist in large proportions, lime and potash; and hence, it must exhaust a soil as much as timothy or any of the best grasses. For this reason, clover makes an excellent green crop to precede wheat. Its large roots loosen and open the soil, and supply by their decay a large amount of fertilizing matter.

I have already remarked that clover has not succeeded well in this State. In many instances it has not come up, and in others it has died out. In some instances it has not been difficult to assign a reason for its failure. Where it has failed to grow, I found on enquiry that it had been ploughed in; buried too deep. The seed, in these cases, was not in fault. Clover requires only a shallow covering, and especially if the roller is employed, good seed will come up. In other cases, after it had come up, the planter allowed his pigs to have the benefit of the young and growing plant. It was, therefore, fed or crushed out. In other cases it was sown at the wrong time and was exposed without protection to the sun-rays.

In nine cases out of ten, a good stand may be secured under the right system of culture. All those causes of failure which I have named must of course be avoided, and in this climate it will not do to allow cattle and hogs to feed upon it until it is half grown, or has acquired a strong root.

For a green crop to be disposed of as a fertilizer, clover has one advantage over the pea; from the former, a good crop of hay may be obtained, and at the same time its stubble and root ploughed in. The latter, if taken off for fodder leaves on the ground only a small remnant of fertilizing matter. But if the whole pea is allowed to remain, it is more valuable than clover, and is better adapted to this climate, and hence requires much less care in its cultivation.

White clover is a more hardy plant than the red, but being much smaller, it is not useful for winter fodder. For fine pastures it is one of the best of plants, though cattle do not relish it quite as well as we have reason to expect from its sweetness and tenderness; yet, is eaten freely by sheep, and the meat, whether of cattle or sheep, is of a fine quality. It is also relished by swine. Its root being creeping, it spreads far and wide, and makes a durable pasture, which bears close feeding remarkably well. Butter and cheese made from the milk of cows whose pastures are dotted with the white clover, is superior to any other, all things being equal.

White clover contains, when fresh and healthy,

Water,	81.50
Dry matter,	16.76
Ash,	1.75

In one ton of clover there are 234.08 lbs. of inorganic matter. The ash I found composed of

Silica,	28.075
Phosphate of lime, magnesia and iron,	19.325
Carbonate of lime,	16.730
Magnesia,	2.175
Potash,	10.880
Sulphuric acid,	2.305
Chlorine,	0.615
Carbonic acid,	4.234
	99.979

The white clover differs from the red in the composition of its ash in containing a much larger amount of silica. It may turn out that the foregoing determination is erroneous or is too large. It may be accounted for, perhaps, by supposing that fine sand adhered to the stem and leaves.

LUCERNE—(MEDICAGO SATIVA.)

§ 114. This plant belongs also to the leguminosæ or pea tribe. It is an inhabitant of a warmer climate than red clover. It has been cultivated for fodder or the food of cattle for twenty-three centuries.

Lucerne requires a soil especially adapted to it; it is not therefore so easily cultivated as clover. It requires a tolerably rich soil, and one that is mellow and permits its roots to penetrate deeply. A light sandy soil does not suit it, neither does a stiff subsoil which retains moisture strongly, or is impervious. A fair proportion of sand, clay and vegetable mould will be found a suitable mixture for the growth of lucerne. The climate of North Carolina is well adapted to its cultivation. It would undoubtedly grow well and vigorously on many of the pocosin soils, whose composition is similar to that of Hyde county, though probably a better drainage may be required. Still, a soil so well adapted to Indian corn may be expected to grow lucerne equally well. It sends down a long tap root, provided with many fibrous off-shoots, which imbibe nutriment from a wide area. Hence its vigor, when well located, and the great amount of food it furnishes. Lucerne continues to produce good crops from 5 to 10 years in succession. Hence its value; when once thoroughly rooted or set, it is as permanent as the best pasture lands. It would seem, if we reason from the effects of the cultivation of other plants, that after 10 years cropping the soil would be perfectly exhausted. This is not the case, for it is said to render the soil richer. This is going too far. For though leguminous plants derive a large portion of their solid matter from the atmosphere, yet the inorganic matter comes from the soil, and just as much of it as is removed from the field, just so much also is the land impoverished. The reason of the anomaly claimed for lucerne, is, that it penetrates much deeper than other plants and takes its food from a much greater space.

The best time for cutting lucerne is just before it blossoms. If cut before this period it is too watery to dry and cure well; if later or after blossoming it is too woody and contains less nutriment. This is probably one of the best plants for soiling cattle. When cut it sprouts vigorously again, and in a climate like that of North Carolina, it seems to be the plant which may be relied upon to stand the sun and drought, and at the same time furnish a forage superior, if any thing, to the red clover. The seed of lucerne are yellow, and if good, glossy and heavy. The first year it should not be cut too close nor a large amount of forage expected from it. Time should be given for it to take deep root. The second year it begins to pay and may be relied upon for several succeeding years. It should be sown early in spring.

According to Prof. Way, the proximate elements of lucerne are as follows:

Water,	69.95
Albuminous matter,	3.83
Fatty do	0.82
Heat-producing matter,	10.32
Woody fibre,	8.74
Ash,	3.04

When the plant is dried in a water bath at 212° Fah., the albuminous matter amounts to 12.76, and the heat-producing to 18.62 per cent. The albuminous matter or flesh-forming elements of the Kentucky blue-grass are 10.35, and its heat-producing matter to 43.06. It is therefore superior in flesh-forming elements to this favorite grass.

SANFOIN—(HEDYSARUM ONOBRYCHIS.)

§ 115. Like the clovers and lucerne, sanfoin is a leguminous plant, but differs from the latter in many important particulars. It has many long leafy stems. The leaflets are smooth and pinnate, or in pairs, rather oblong and pointed, and slightly hairy on the under side. Flower stalks are terminal and extend above the leaf stalks, and arranged in the form of a spike, with crimson and variegated blossoms. The stems grow from two to three feet high. The pods are flat, hard and toothed on the edge; root perennial and woody; flowers in July.

According to the opinion of an experienced English agriculturist, who has resided many years in this country, the sanfoin will prove a valuable addition to the artificial grasses of this country. The following remarks containing a summary of his opinions I propose to embody for the consideration of the planters and farmers of this State.

In the first place, it will grow well on light soils, sandy and gravelly loams. It may be sown after rye or barley, and should not be fed the first year, or immediately after the crop is removed. It may also be sown with grass seed. The following year it may be mowed, and then it is in a condition to be fed by sheep.

This plant is probably better adapted to horses than cattle, especially milch cows, or rather horses and sheep. Sheep consume the leaves and softer parts of the stems, and then horses eat readily the remainder. Working horses do well on what sheep leave. Sanfoin has been mown for nine or ten years in succession, and has produced good crops each year without manure. It is not the proper food for milch cows, as it imparts a bitter taste to the butter. The sod, after it has been growing for several years, is full of roots, and it is often ploughed and then burnt over. In this climate ploughing and burning is not advisable.

The nutritive value of sanfoin does not differ materially from lucerne. It is composed, so far as its proximate elements are concerned, of:

Water,	76.64
Alluminous matter,	4.32
Fatty matter,	0.70
Heat producing,	10.73
Woody fibre,	5.77
Ash,	1.84

When dry, it yields of alluminous matter, 18.45, and heat producing, 45.96.

CRIMSON CLOVER—(TRIFOLIUM INCARNATUM.)

In some parts of this country this clover would no doubt succeed. It however, requires a climate rather cooler and moister than that of the eastern counties. But in the mountainous section of the Southern States it can hardly fail of being received with favor. The advantages arising from its culture, are, it may be sown after potatoes are secured, and produce a spring crop which will be earlier by eight or ten days than lucerne or red clover. It produces two good crops in one year. It is, however, an annual, and it requires as much care to insure success as the red clover. For soiling cattle it is well adapted, in consequence of its early growth. If cut for hay, it should be gathered as soon as it is in flower. The seed may be obtained from the second crop. As a general rule, where the red clover succeeds, it may also be expected that the crimson clover will succeed also.

CHAPTER XIV.

Methods by which the valuable grasses may be cultivated successfully—Soiling, and its advantages.

§ 116. In this State it is important in the first place to select the proper field for the cultivation of grass which it is designed to cut for winter fodder. It appears to the writer that as summer heat and drouth are the greatest obstacles to the successful cultivation of grass and hay, that such fields should be selected as suffer the least from the operation of these causes. Hence it is believed that the meadows and low grounds which are bordered by permanent streams and which are naturally quite wet, but may be laid comparatively dry are the most suitable for grass lands. The first work which is required, is to drain the field thoroughly by ditching. Fields of this description are invariably supplied with a rich bottom, which is capable of furnishing an indefinite amount of nutriment, or sufficient to sustain crops of hay for years in succession, and being also supplied with water which percolates through the lower strata of earth, are little liable to suffer from summer droughts. Besides, these low, flat meadows may be cheaply irrigated if necessary. Irrigation is also one of the cheapest and most effectual means by which nutriment may be conveyed to the grass. The great object, however, to be attained in the selection of such field, is that of securing a cool and moist soil, for many of the best grasses are found flourishing under those conditions, though they by no means grow in wet bogs or swamps. Timothy, one of the best of the Northern grasses, grows best in a moist soil.

After a drainage has been effected, many of the wild and least useful grasses will die out. But to aid the process of substitution of better, for the poorer grasses, and the weeds which always, more or less, take a joint possession of such fields, it may be harrowed with an instrument provided with sharp teeth. When this is done, a proper mixture of seed may be sown, after which the surface is swept over with a heavy brush.

The introduction of the valuable grasses is also materially aided by a top dressing of compost, which puts the soil in a better condition to receive the seed, and facilitates, as well as quickens, its germination. It also gives more strength to the newly introduced grass, and enables it to contend more successfully with those which are already in possession of the premises. As in law, so in agriculture, possession gives important advantages; and the new claimant which we desire to put in possession, must, in the first place, oust the old occupant. Much depends upon the perfection of our preliminary steps. If we have thoroughly under-drained the premises, we shall be enabled to starve out very speedily the occupant we wish to remove; and if, in addition to this, we supply nutriment to our favorite intruder, we have provided or opened more than one way by which we hope to succeed. The poor grasses are generally destroyed by high cultivation, and so are weeds, and the process which so evidently favors the disappearance of the poorer ones, favors the introduction of the good. One of the most substantial reasons why grasses are so difficult to grow in the South, is, that they are not manured. They are sown first upon soil already partially exhausted, where the poor grasses are taking deep root, and hence their chance for life is very small.

If a grass plat is to be formed upon upland, the proceeding should be somewhat different. After the land is made even by light ploughing and harrowing, winter rye should be sown, and the field stocked down with orchard grass, mixed with herds grass, june grass and red and white clover. The rye makes an excellent spring fodder, and protects the grass seed, which in due time will germinate and replace the rye. To insure success, let the seed be sown thickly, not sparingly, for the writer believes that in the climate of North-Carolina more seed is required than where the climate is cooler. Besides, there is no more effectual means to guard against drouth, and a hot sun, than to cover the whole surface with vegetation, and the supplying this vegetation with abundant nutriment. In support of this view, let a field of Indian corn be examined, a part of which has grown sufficiently to shade the soil, and part is backward from any cause, and does not shade it. The first will sustain a drought without material injury, while the other will be destroyed. So also, where clover has taken a strong and vigorous hold and covers the ground, it stands a severe drouth, while that portion of the field which is thinly planted, dries; the soil becomes hard and cracks, and the plants perish. We may, therefore, be guided to successful results by observation. What frequently takes place naturally, or accidentally, in consequence of a failure in our own experiments, will furnish safe ground to go upon. We cannot insist too strongly in this climate upon the use of much seed, that the soil may be covered with vegetation; and hence, protect it by preserving the surface in a cool condition. Moisture is always condensed from the atmosphere upon such a surface during the night, and evaporation is in a great measure prevented by day, if a thick coating of vegetation has grown upon it. We should not forget in this connexion that early planting is one of the means by which we may secure a crop from the effects of a drouth.

One of the best materials for grass lands is ashes, either leached or unleached. The latter will, of course, contain less potash, but even then, they are highly valuable. In the absence of ashes, fine marl sown broadcast, or if accessible, strewed freely upon the surface, will effect important results, either ash or marl bring in clover, without sowing seed. Plaster produces the same effects. Where a system of husbandry is pursued which furnishes barn-yard manure, it supplies an admirable basis for composting. Very few plantations in the eastern section of the State, which do not furnish muck or peat. With one load of barn-yard manure and two loads of muck or peat, three loads of an excellent fertilizer may be made. These materials should be well incorporated and receive from time to time all the refuse matter of the house, yard and garden, or anything which will ferment under the influence of the necessary conditions. Wool, hair, refuse animal matter of all kinds, become of the utmost importance in composting. One important addition should not be neglected; that is plaster of paris. In the absence of that, dirt sprinkled with copperas water, which is not expensive, will make an absorbent of the gasses. That dirt alone, or earth, has strong absorbent powers, we have sufficient evidence in the fact, that very little odor escapes from the carcass of a decaying animal body when it is perfectly covered. But additional earth should be added from time to time, as the first becomes saturated with the effluvia. The matter which escapes under these circumstances, is ammonia, which is one of the active principles paid for in guano, which makes the difference in the price of Peruvian and Mexican guano. Compost heaps require a small proportion of lime, but wherever animal matters or excrements are concerned, there should be a large intermixture of muck or peat. No good farmer adds lime to his barn-yard manures; it may be done only where undecomposed vegetable matter is ready to absorb the disengaged ammonia.

SUMMER SOILING.

One of the most important measures for carrying on a successful and profitable scheme of husbandry, is to incorporate with the general plan or system, that of soiling cattle. Its value has been fully established, both in this country and Europe. Apparently, it is objectionable from the amount of labor it requires; but this objection vanishes when it is put in practice, and becomes the every-day business of those appointed to superintend it. Cattle, when soiled, must be confined to a yard, at least, and fed on mown grass, lucerne, clover, or corn sown broadcast. A large stock may be kept on five acres of ground, or, it may be made to yield that of thirty acres of pasture lands. After being fed in stables, they may be driven to a pasture for the purpose of exercise, and returned again at night, and fed on fresh mown fodder in the morning. Soiling is no doubt as well adapted to the South as in the North. By this system, cattle are protected from a burning sun during the day,—a protection which is almost as important as protecting them from the cold. Most farmers appear to forget that good stock are like the cereals, which have been brought to their best and improved condition by artificial means, and the moment the efforts to maintain them in this highly improved state are suspended, they begin to deteriorate. Cattle can no more be kept in a good and prosperous state than the cereals, which if the condition of the soil is neglected, fail to produce remunerating crops. But furnish them with food and place them in comfortable circumstances, and profits are sure to be returned.

Soiling is adapted to the circumstances attending the cultivation of a few or many acres. The system consists in cultivating those grasses which come to maturity in succession, and it is desirable to be able to vary the kinds of green food every few days, though it is not necessary to the success of the system.

In connexion with summer feed, it is important also to have an eye to the winter support of the same herd. For this purpose root crops become an important part of the system of soiling. When, for example, the patches of corn, oats or rye are cut up, the sugar beet or turnip may be sown for winter feed. To these, then, should be added carrots and sugar parsnips. The object of root culture for stock is to supply a variety of nutriment for horses and cattle, which, if fed with them once a day, may become much more thrifty and healthy than if fed only upon dry fodder. For a Southern grass, the orchard grass should take the place of Timothy. This, with the June grass, red top, and herds grass, and a few others already described, will give all the winter hay which may be required. The practice of *pulling fodder* from the Indian corn is much more laborious and attended with more trouble than that of mowing grass for hay. An acre of sugar beet will produce a thousand bushels, and an acre of carrots over six hundred, and the sugar parsnips yields about eight hundred bushels to the acre.

One of the incidental advantages of soiling is the production of a large amount of valuable manure which may be saved under cover, and to which may be added the refuse of the kitchen and garden, whereby its quantity may be indefinitely increased.

In the foregoing observations upon soiling, I have been disposed merely to allude to the subject, believing that those planters who wish to keep good stock, either of horses or cattle, will be inclined to try this as a part of their system of husbandry; a system, which, if carried out, will not fail to give them a good stock of cattle and cows as well as horses, all of which may be kept cheaper and better than in the mode now pursued in this State.

CHAPTER XV.

PALAEONTOLOGY.

Fossils of the Green Sand and Tertiary—Mammals—Horse—Hog—Mastodon and Elephant—Deer—Whales, or Cetaceans.

The distinguishing features or characteristics of any age or epoch, can be known only from the history of the men who were then living. The characteristics of the age when the Romans were gaining an ascendancy in the world, can only be known from the individual or collective memories of Roman citizens. A history competent to give us a knowledge of those times, would blend together the personal appearance of men, their habits, dress, food, etc., from which we should also obtain facts or inferences respecting the country, its animals and plants, its climate, topography and grand divisions. So of Greece, Egypt and Palestine. The memories of the actions of these nations in their generations, would furnish us the leading facts respecting the characteristics of the period in which the respective nations lived.

So, also, the characteristics of the fossils furnish at least a clue to the features of the epoch during which they lived. To determine these features, demands an intimate knowledge of the present; for, we are under the necessity of comparing the past with the present. The present is the standard, and no comparison can be made of any value which neglects the present. We find in the present certain structures and forms which we know have certain relations to climate, or to the conditions in which they exist. If, then, similar structures or forms are found attached to an extinct being of any epoch, it is a fair inference that that structure or form bore a similar relation to the external conditions which surrounded it. Its full description, then, would be a memoir of the animal, its habits would be indicated, its relation to surrounding circumstances would be known; many inferences would follow from each,—some would bear only upon its instincts, its food, its means of defence from the medium in which it lived, etc.

If, for example, an oval shaped bag filled with coloring matter, in connection with a fossil known as the Belemnite, it would be inferred that this bag contained a fluid designed to conceal it from its enemies; that it would deeply discolor the water into which it was cast, and thereby, under its cloud of dye-stuff, make its escape. Such a phenomenon is familiar now to the sailor. The cuttle-fish is thus supplied with dye-stuff, and he employs it for escaping from a pursuing enemy; and as this is so, so it is inferred, the animal did which was supplied with a similar apparatus in the period of the Lias and Chalk.

We might go on and note hundreds of analogous examples, but one must suffice. This view is borne out by one great and leading fact, that all extinct animals are constructed upon one of the four leading types which now prevail. Of the millions of individual fossils which have been seen, not one is known which does not belong to, and may be referred with certainty, to one of the great leading types of the present. It is the plan then, which really tells all this, or makes it possible to compare and infer with certainty. Observation is the way, but the plan of creation makes it possible to deduce a connected history of the past from the dead races, and thereby see at a glance how any former epoch differed from the present, or from those ancient ones with which it was more intimately connected.

My object, however, is not so much to direct the student in this chain of reasoning, or so to apply knowledge as to make him acquainted with the external forms of the fossils of the marl beds. The figures and descriptions will enable him to know the objects from their forms, and thereby to distinguish the marl beds which contain them from each other. It is, therefore, a practical subject, and may be studied as such. But the knowledge thus acquired prepares the way for further advances in science.

The fossils described in this part of the Report, belong to four or five periods, inasmuch as some of them are found in two or more successive ones. These periods have been distinguished by the following names which are expressive of certain ideas. Thus, the oldest is the cretáceous or chalk formation. It is, however, only a small part of it, and that part is the inferior or oldest part of the cretaceous system. This part is widely known as the Green Sand, and has been employed extensively as a fertilizer. The 2d, in the ascending order, is the Eocene, which means the dawn of the present, as a few species survive, which were created in this epoch or period. Only about four per cent., however, have lived on through all the vicissitudes of the times. The third, is the Miocene. Of the animals created during this period, more than half have perished, and we know them only through their remains. The fourth is the Pliocene,' the animals of which less than half have perished. The fifth, the post-Pliocene, is known by its fossils being similar to those which now live, excepting five or six per cent. Hence, it may happen that one of the four species of animals which survive, and which was created in the Eocene period, may be found in all the succeeding beds, but it is evident it will be associated in each case with races or species quite different from those among whom it was first connected or who were its cotemporaries.

The cause of the extinction of so many species, is a mystery. The fact is well established, but it is only in certain cases that we can account for their disappearance. It appears to have been sometimes due to a sudden catastrophe, the ejection of mud, or poisonous matter into the medium in which they live. This happens now, and probably has happened before, but in a majority of instances, it is impossible to perceive any external cause which destroyed them; and hence, we are left to speculate on probabilities, without being able to arrive at satisfactory conclusions.

MAMMALIA.—EQUUS CABALLUS.

There is scarcely a question so interesting to the naturalist and historian as that which relates to fossil remains of the horse. The testimony of historians is, that the horse was not living upon this continent at the time of its discovery by Columbus. The testimony of the naturalist is, that the horse lived upon this continent at a period prior to its discovery, its remains having been found first in the miocene, and lastly in the pliocene, in which period it may have become extinct. Its earliest appearance is in the former; and it appears from the discovery of Prof. Holmes, of Charleston, S. C., that its remains are not uncommon in the latter.

FIG. 18.

Figure 18 represents the crown of the third or fourth molar of the left side of the upper jaw. It has complicated enamel plates, or columns, and is somewhat worn, but by no means an old tooth, as its roots are undeveloped. It is two inches long and an inch thick. It is undistinguishable from the corresponding tooth of the recent domestic horse. It is a deep brown color, and looks like a fossil.

Figure 19 represents the crown of a tooth of the third or fourth molar, probably the third, of the left upper side. It has not been worn. It resembles a recent tooth, as it is whitish, and only stained brown on one side. The enamel plates, it will be perceived differ from the preceding, and they differ also from those of the corresponding tooth of the domestic horse. This difference, however, may arise from its unworn condition, as the enamel plates differ somewhat in configuration as they wear down. This tooth is three inches long and one thick.

Fig. 19.

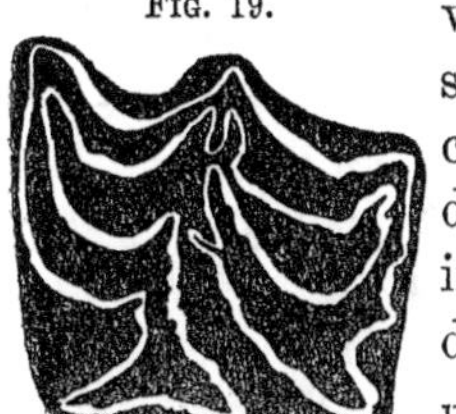

Fig. 20.

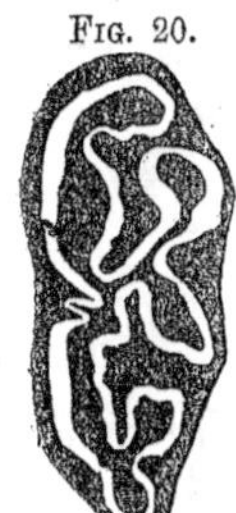

This figure (20) represents the back molar of the left side of the lower jaw of the horse. It differs only slightly from the corresponding tooth of the domestic horse. It is worn, but belonged to a young individual, and its roots are undeveloped. It is three inches long, one-half an inch thick, and one and a quarter wide.

Figure 21 represents one of the incisors of the horse; a, front side; b, inner side; c, lateral view. This scarcely differs from the corresponding incisors of the domestic horse.

Fig. 21.

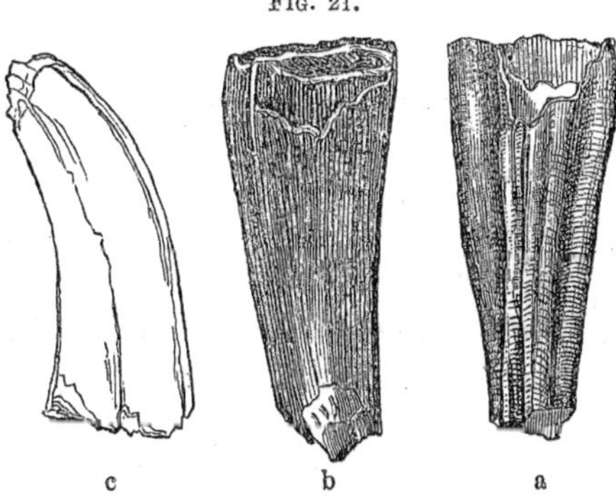

c b a

The foregoing teeth are from the miocene of North-Carolina, and were discovered at an early period of the survey. No. 18 was found in a bed at Elizabethtown, Bladen county, and was accompanied with a tooth from the lower jaw. No. 19 and 20 are teeth washed up on the beach at Plymouth, N. C., and

No. 21 from the miocene of Pitt county. I found, also, molars, in Pitt county. They occur in a sandy bed, which may be ten or twelve feet above the shell marl. Although there is a close correspondence between the fossil teeth above described and those of the domestic horse, which was introduced into this country since its discovery, still, it is probable that it is a different species. If it is maintained that the fossil and introduced species are identical and the same, it follows that the same species was created about the same epoch, in two very different quarters of the globe, viz: Asia and America, and in climates which differed materially from each other. Farther discoveries must be made before this interesting question can be satisfactorily settled.

SUS SCROFA.—HOG.—(Fig. 22.)

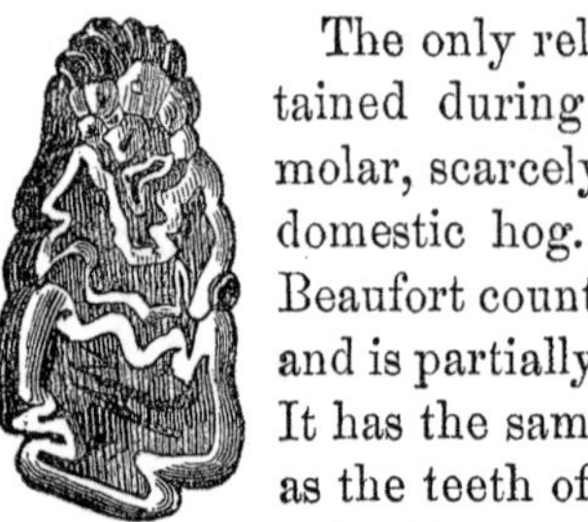

(Fig. 22.)

The only relic of the hog which has been obtained during the survey, is the last inferior molar, scarcely differing from its fellow in the domestic hog. I obtained it at Washington, Beaufort county, from the miocene. It is brown, and is partially mineralized by sulphuret of iron. It has the same claim to genuineness as a fossil, as the teeth of the horse already described.

The hog was introduced into this country at the time of its settlement, but as in the case of the horse, it was peopled by this interesting animal a long time prior to its discovery. It also became extinct, and at its settlement was supplied again from a foreign country.

PROBOSCEDIANS.—MASTODON GIGANTEUS.—(Fig. 23.)

The bones of this large pachyderm are not uncommon in the miocene marl of North-Carolina.

Fragments of ribs and bones of the extremities are the most common. The figure of the superior part of the crown in the margin was taken from a tooth found in Halifax county. Its enamel is jet black and highly polished. It is the first or small molar of the right side of the under jaw. It is an old tooth with the lubercles worn down, and was probably

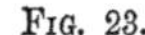

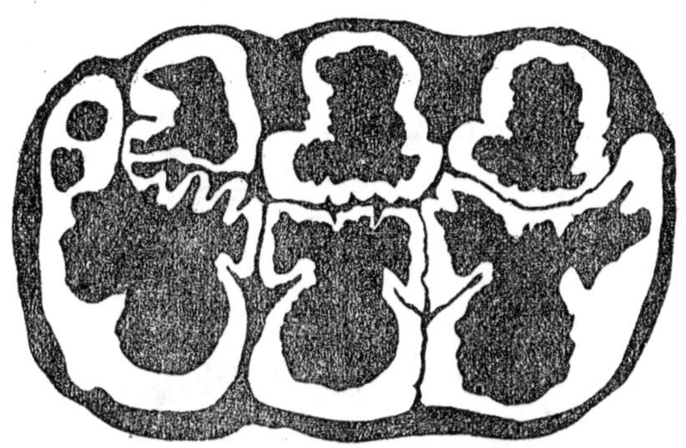

lost or shed while the animal was living. The figure is designed to show the arrangement of the enamel plates.

Bones of this immense quadruped have been found at numerous places. A large number were found in a marl pit near Goldsboro', and a large back molar in another marl pit in Nash. These bones are usually broken, and the pieces are rarely more than from three to six inches long. A cuneiform bone of the foot was found in a marl bed upon the Cape Fear. From the number of bones which have been found it is evident this large species of land quadruped, the largest known, must have been very numerous at one time. Its bones are associated with fossils, many of which are now extinct, and some or even many still survive. The oldest deposit in which the bones of the mastodon are known to occur is probably the miocene. They continued to occur in the subsequent formations until the latest, which just precede the advent of man; and, indeed, it is not at all improbable that man witnessed the final extinction of the race. The long bones which I have examined always contain animal matter, an evidence of their recent death.

The elephant was also a cotemporary with the mastodon. No teeth, however, have yet been found in North-Carolina which may have enabled me to identify its remains. But to those who have marl beds to identify its remains, a tooth (Fig. 24) of this interesting animal is given in the margin. It is a reduced figure of one found in the superficial deposits of New

York. A tooth belonging to the elephant was taken from the beach upon Seneca lake New York, and portions of a skeleton were found near the surface in Monroe county. All these bones contain also animal matter, and they are usually associated with moluscous animals which are living at the present time.

Fig. 24.

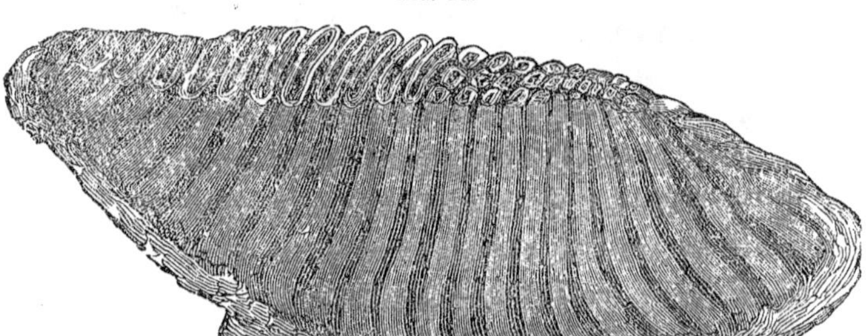

It is probable the mastodon lived in a period prior to that of the elephant, but it appears that both became extinct at or about the same time.

That the mastodon and elephant roamed in herds over a large part of this continent, seems to be indicated by the fact that their bones are found from the Atlantic to the base of the Rocky mountains. The bones of the mastodon, however, are more numerous and more widely extended than those of the elephant.

RUMINANTIA.—CERVUS VIRGINIANA

The discovery of the remains of the C. Virginiana deer, is an interesting fact. It appears to have been cotemporary with the Mastodon and Elephant, which have become extinct. So, also, it is cotemporary with the great Irish Elk, which has become extinct in Europe.

The base of the horn which I found in the Miocene bed about 10 miles above Elizabeth, on the Cape Fear, is about six inches long. In this horn, the first branch goes off from the axis nearer the head than usual, but this occurs occasionally in individuals of this species.

It appears from this discovery that the common red deer of America began its existence at or about the same period as the American horse; but while the horse became extinct, the deer has survived. In a fresh water marl bed, in Orange county, in New York, I found a horn of an extinct deer which was associated with the remains of the mastodon. The deer of the miocene marl survives, while a more recent species has become extinct, or such is the evidence of facts as they now stand.

REMAINS OF THE PORPOISE.

Several vertebræ which appear to have belonged to the porpoise, have been obtained from the marl beds near Rocky Mount. They appear to belong to a species which differs from the common one of the coast. The figure shows the end of the vertebræ to which the intervertebral substance is strongly attached; the other extremity is smooth. The body is encircled in part with a deep channel or groove, which is connected with the holes which transmit the vessels, and nerves at the base of the spinal arch.

In addition to the foregoing remains of the order, cetacea, I may mention the occurrence of the Zeuglodon cetoides. (Owen,) a fossil of the eocene, which was first found in Alabama, and described by the late Dr. Harlan, of Philadelphia. The teeth are entirely unlike those of the common cetaceans, and belong to a type not very unlike those of the seal. No teeth, however, have as yet been discovered in this State.—The remains of this cetacean consist of vertebra which were obtained from Washington, near the line of the Wilmington Rail Road.

One of the largest candal vertebræ of a whale, (fig. 25,) has broad flat transverse processes, standing at right angles to the body of the bone, the articular ends are unequal, the anterior being 5½ and the posterior 4¾ inches in diameter, and circular, with a length of 6 inches. Of this length the base of the transverse processes occupies 4 inches, and terminate behind in a rounded notch; their length is 2½ inches.

FIG. 25.

LOWER JAW OF A BALAENA OR WHALE.

On the Meherrin, near Murfreesborough, I found portions of three lower jaw-bones belonging to the genus Balaena, together with many vertebræ, all of which appear to belong to one species.

These jaws are imperfect,—the anterior part the left lower jaw is smooth, gently covex, and curved on the outside, but rather flat inside. The wide upper margin is perforated with three holes penetrating the jaw in a slightly descending course, and terminating anteriorly in an edge produced by a champering of the inside extremity, and rounded from the base up to the upper edge, which is grooved for six inches. They are 3¼ inches wide and 2 inches thick, and nearly straight. All the posterior parts of the jaw had been lost, and only two feet obtained. It is impossible to refer these fragments of jaws to either species which furnished the ear bones, as neither of these specimens were obtained at this locality. But the vertebræ and jaws belonged to one species, and it is possible hereafter to determine to which ear-bone belonged to the Murfreesborough species. It is evident that neither of these belonged to Prof. Leidy's Orycterocetus, because this belonged to a different family of the cetaceans.

OTOLITES, OR THE EAR BONES OF WHALES.

The remains of the cetacea may be said to be numerous in the miocene of North-Carolina. Vertebra and ribs are more commonly found than other parts for the reason that the individual parts exceed in number the other parts of the skeleton. The ear bones are the least common. Of this part I have those which I regard as having belonged to at least three different species. I base this conclusion on the established fact that these bones possess for each species a peculiar configuration; that though the bone in question has a general resemblance in all the species of which the family is composed, yet in the minute details of construction and form, each species has its own, which may be determined by close and careful comparisons. Thus, in the true whales, the thick posterior part is simple, while in the cachalot it is bilobed, and that this thickened and convex part in the simple kinds, while it is variable in form and extent in the different species of the true whales, and which is also joined to certain other differences, which may be observed in the thin overarching and expanded part.

For convenience of description, these bones may be divided, longitudinally, into two principal parts: 1. The thick involuted convex part which occupies the posterior segment of the bone, and which extends back to a rough longitudinal surface; and, 2d. The thinner and expanded part which begins where the former ends, and arches over the first in different degrees, forming, posteriorily, a convex surface, and interiorly towards the first part a concavity differing both in degree and extent in different species. The anterior or eustachian portion is formed wholly of the thinner expanded part. There is in the form of the expanded part some resemblance to the rim of the human ear.

The ear bones, in consequence of the thick convex part

being simple, are all referred to the genus balaena. Other parts of the skeleton of this genus have been formed, as the vertebrae, ribs, lower jaw, &c.

The first of the bones (Fig. 26) which I propose to describe is the largest, and resembles in form the same bone belonging to the right whale, (the balaena mysticetus.)

Fig. 26.

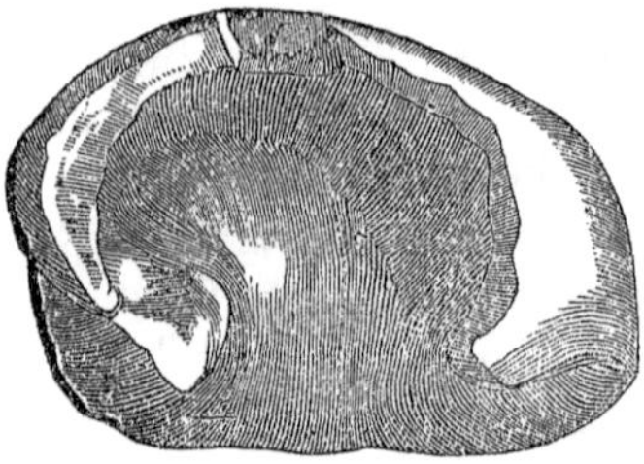

In this specimen the thick involuted part is thickest at its extreme posterior end, and gradually diminishes to within three fourths of an inch of the flatish, expanded or eustachian part of the tube.

Its surface, as it passes backward, and corresponding to the span between the lobes in the cachalot, becomes slightly concave, and the whole surface to the boundary backwards and forwards to the channel, which separates it from the concave expanded portion, is irregularly wrinkled; these wrinkles increase in strength to its junction, with the latter part, where the line of division is distinctly defined. At the posterior part, there is a strong indentation, somewhat in the form of the letter U, surrounding the part where the expanded part springs. The thinner expanded part forms an arch, concave within, and quite regularly convex without; at the extremities it forms expanded hooks. The concave surface widens from the posterior to the anterior end, and is widest just within the margin. This bone differs from the same in the right whale, in its convex portion being lower and not above the level of the concave cavity beneath the arch; and being, also, perfectly separated by a change in the appearance of the part, and also by the perfect smoothness of the concave surface of the overarching wall, which, in this B mysticetus, is very rugged.

Its length is 3¾ inches, and width 2¼, and belonged to a large whale, though probably not the largest. It is, however, very bulky. Cuvier remarks, that the ear bones of the Balaeonoptera are very small in proportion to the size of the species; so that it does not follow that where the bone is small the spieces must be small also.

I propose the name *Balaena Mysticetoides* for their species.

The thick, the posterior end, is nearly equally bisected by the thin expanded part, and around it there is a deep sinuous indentation which, on the inside, is continuous with the channel between the thick and thin parts.

Fig. 27.

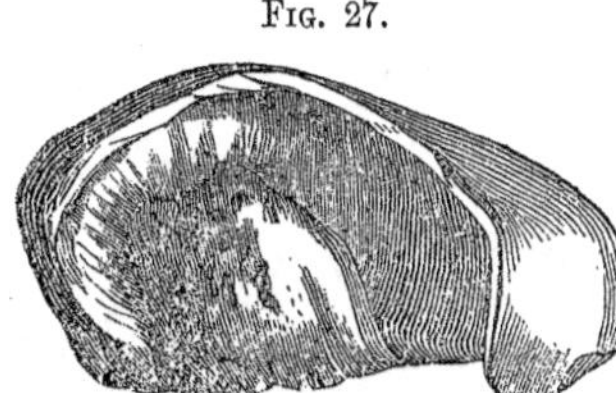

The otololite, next in size to the B misticetoides, differs much from it in form and proportion of parts. The thick convex part is well defined, but rough, short and prominent. It rises higher than the base of the thin involuted part to which it slopes all round. It is marked with two or three strong folds, one of which is at or near its termination forward, and another beneath, which gives a slight emargination to the bone. It is separated fromthe anterior end by a flattened plane about half an inch wide, where their expanded part turns and forms a rather open hook, unlike that of the former, which is bent much more inwards. The posterior end is somewhat obliquely truncate, and at the root of the thin part there is a rough indentation disconnected with the wide channel within. The anterior border of the thin part forms an arch much less extended than the former, and the posterior and basal part is flattened and angular. Length $3\frac{1}{2}$ inches; widest part $1\frac{1}{8}$.

Another specimen measuring four inches long preserves the essential characters of the foregoing. It is very rugose around the thick convex part, and the middle fold creates a slight twolobed character to the interior part and its base.

The smallest (Fig. 28) has a well-defined convex part, which is smooth though somewhat wrinkled, but rough within, and the border rises almost immediately from it, especially posteriorly. The space between the border and convex part widens anteriorly where it is only gently

Fig. 28.

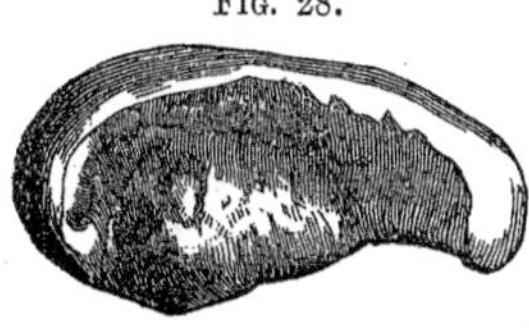

curved, scarcely forming a hook. Behind the convex part it is very regular, but the beginning of the thinner expanded part is formed by a rounded ridge which may be traced from one extremity to the other. It is far less angular, and more regular than the preceding. It is 2¾ inches long; greatest width 1½ inches.

This ototite is one of the most common in the miocene beds. Unfortunately, in all these specimens, the thin expanded over-arching part is broken off, but it is evident that in this case this part was very limited.

The two smallest are perforated by boring moluscks, a fact which shows that instinct is sometimes at fault.

It is probably impossible in the present state of our knowledge of the anatomy of those extinct whales, to refer them to the species to which they belonged. That the foregoing ear-bones I have described belonged to different species of the whale, there can be no doubt.

Few extinct species of balaena are known to belong to the miocene period besides the orycterocetus of Leidy.

SUMMARY

Of the characteristics of the three foregoing species, derived from a comparison with each other, and with the three which have been described, by Prof. Owen.

The B. mysticetoides differs from B. affinis Owen, in the much greater extent of the overarching wall and the well defined limits, and greater prominence of the involuted part;—this part also bears a much greater proportion to the whole of the organ than it does in the affinis.

The B. definita Owen is very strikingly truncated at its posterior end, and has also its thick involuted part much less in proportion than in the B. mysticetoides, and its thin overarching border is also much less in extent.

It differs from the B. gibbosa, Owen, in most of the characters just stated; particularly the extent of the overarching wall, its thick convex part is much less prominent; but it resembles the B. gibbosa somewhat in its configuration at the posterior end, where the rim is continued around it, as it were, but in the gibbosa, it rises from near the base, while in the *mysticetoides* it rises higher and is surrounded by deep sinuous indentations. It resembles also the B. *emarginata* in the existence of a concavity on the inferior border of the thick convex part, but is much less; the overarching wall exceeds very much in extent that of the *emarginata.*

The figure 27 differs from the affinis in its prominent and distinctly defined convex involution. It resembles the B. definita somewhat, in its posterior truncation ; but the involuted part is more prominent, and has a strong ridge or prominence on the border near its slope to the concavity ; but it resembles still more closely the B. gibbosa, in the form of the convex part, but the thinner overarching wall is more extensive and broader at the eustachian termination, and the shape of the posterior end differs from it materially, particularly in the strong angle of the extreme of the overarching wall.

It differs from the B. emarginata, in having a prominence at the base of the involuted thick part instead of an emargination.

The figure 28 differs from the B. affinis in its prominent involuted part, and distinct form or separation from the concave overarching part; from the B. definita by its prolonged posterior part, in which respect it differs also from the gibbosa and from emarginata by its absence of this particular character, and by the presence of strong sugar upon the part next the concavity.

CHARACTERISTICS OF THE EAR-BONE OF THE COMMON WHALE OF THE COAST.

The ear-bone of the Balena Mystictus, the common whale of the coast, in my possession, measured, rather diagonally over the thick convoluted part, is 5½ inches long; the greatest thickness is 3 inches and 3 tenths; the depth or height of the convoluted part is 3 inches; greatest height measured to the top of the thin convolution 4 inches and 4 tenths. The thin involuted expansive is arched so as to have a distance of only half an inch from the thick involuted part. This may be divided into three principal lobes; two of them make up two-thirds of thin part, and these are divided externally by a deep sulcus, and internally by a thick rounded ridge which extends nearly to the base; the lobe of the thickest end is short. A deep sulcated cavity is formed by the thick and thin involuted parts of the bone. This cavity is 3 inches and six-tenths long and 2 inches and one-tenth, and the height nearly 3 inches.

An ear bone having the form and proportions of the Balaena Mysticetus, in many particulars, I have obtained from Craven county. The most important difference is in the height of the thick involuted part, the thin expanded part is broken off but there are so many points of resemblance, that it is highly probable it belonged to this species of whale. The fossil ear-bone is smaller. Its greatest length is only 4 inches and 2 tenths, and the height of the thick involuted part is only 2 inches and 2 tenths. Still, it is not at all improbable that we may regard it as having belonged to the young of the B. mysticetus, and if so this species commenced its existence in the Miocene period. This conclusion is founded upon the probability, that this ear-bone and certain thick heavy ribs of a whale, often found in the miocene deposits, belonged to this species. It is probable, too, that ear-bones vary somewhat in form and thickness in the same species; this is certainly true in the case of the ear-bone of fishes, of which I have many specimens, among which there are several varieties of form and size.

Other forms of cetacean ear-bones occur abundantly in the miocene of Tar River. Figure 28 belongs to one of the rarer forms of ear-bones. It has a distinct involuted portion. It is figured of the natural size.

FIG. 28. (A.)

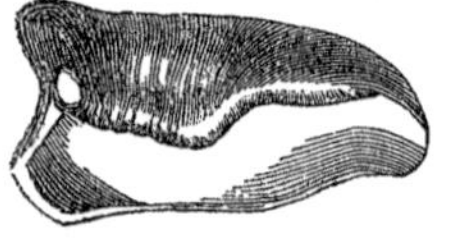

Figure 29 is another form of ear bone which is the most common of all, except the following. It has no distinct invo-

Fig. 29.

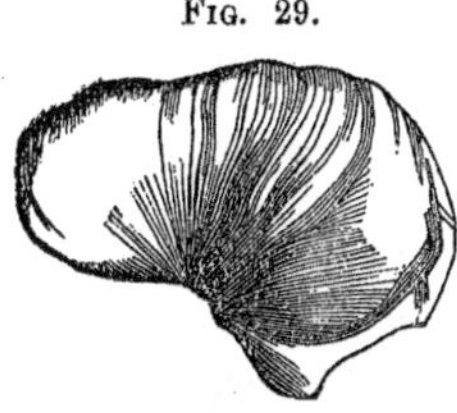

voluted part, though it is thickened at one end of it. It is more or less wrinkled transversely. In other respects it is rather discoidal.

Figure 30, it differs in form from all the preceeding. It is conical, and acute at one extremity and obtuse at the other. From the obtuse extremity, it sends off a short process at right angles, and is probably the point by which it is attached to the interior of the tympanic cavity.

Fig. 30.

But one of the most extraordinary of the ear-bones of this formation, is represented by figure 31. It consists of two parts, a short obtuse conical portion, and a long process extending at right angles from it. It is over seven inches long. The process referred to is four, measured from the base of the heavy conical part, and it extends half way across

Fig. 31.

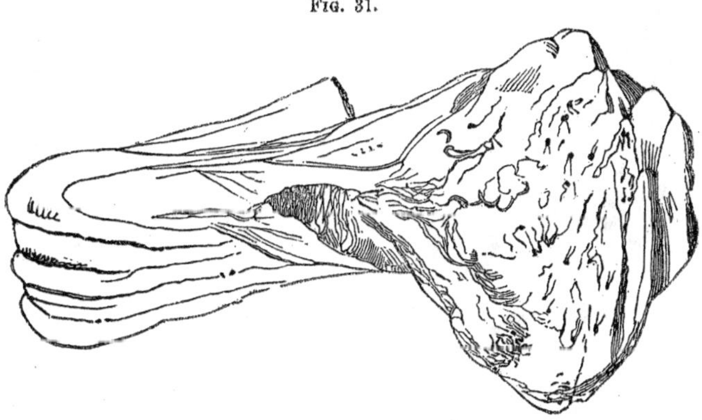

it, so that its whole length is about 5½ inches. The height of the conical part is 3½ inches. This is flattened, and its greater circumference is 8 inches. The arm or process is irregularly triangular, being hollowed out on two sides and flattened on the other. The whole form, however, is difficult to represent by means of a single figure. The figure is one-half the size of the original.

ORYCTEROCETUS QUADRATIDENS.—LEIDY. PROC. ACAD. NAT. SCI. VII, 378.

Fig. 32.

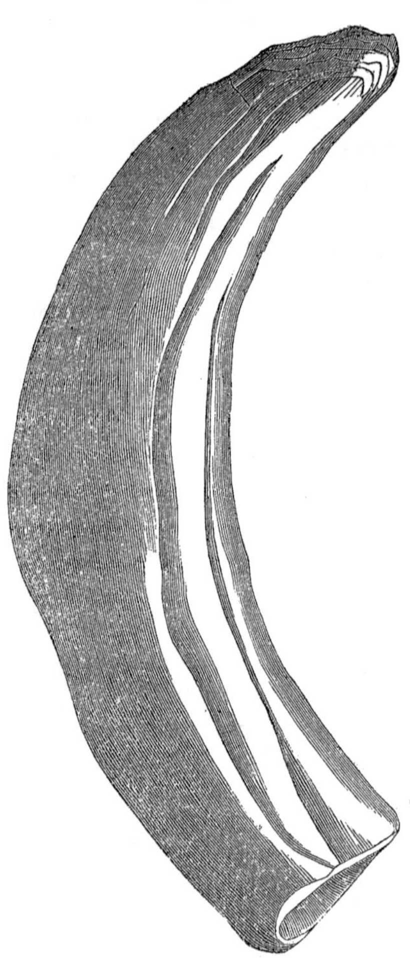

A single tooth belonging to this cetacean was found in Pitt county by Thos. Sparrow, Esq., to whom I am indebted for an opportunity for describing this interesting relic.

The tooth is remarkably curved for a cetacean. It is rather rough, and is somewhat quadrate or angular. This character, according to Prof. Leidy, is not constant. Its transverse section is rather ovate, with the anterior part flattened. It was pointed, but by exposure the apex is injured. Its base has a short conical pulp cavity, less than one inch in depth. Its surface is marked by longitudinal cracks.—The tooth belongs to the right lower jaw, and is drawn the natural size.

It is supposed to belong to the miocene, but the locality contains a few small fossils, derived from the eocene, and hence this may be of that age.

ORYCTEROCETUS CORNUTIDENS.—LEIDY.

FIG. 33.

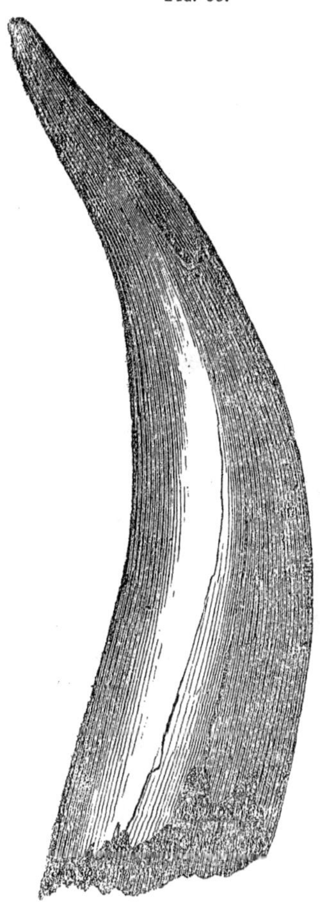

The genus *Orycterocetus* was originally proposed on the fragment of a jaw, and several teeth from the miocene deposit of Virginia. In my collection I have a tooth like those just mentioned, except that it is not quadrate, which it is suspected, however, to be an unimportant character.—The specimen was discovered in the miocene deposit of North-Carolina. It is remarkable for its resemblance in form to a small ox-horn, being elongated, conical and curved. The base is excavated as in the teeth of the spermaceti whale, to which the extinct cetacean was probably allied. In structure, the tooth appears to be wholly composed of dentine. The length of the specimen in the curve is 4½ inches, but it appears when entire, to have been half an inch longer. The section of the base is oval, and is 14 lines in one diameter and 12 lines in the other.

Fig 34.

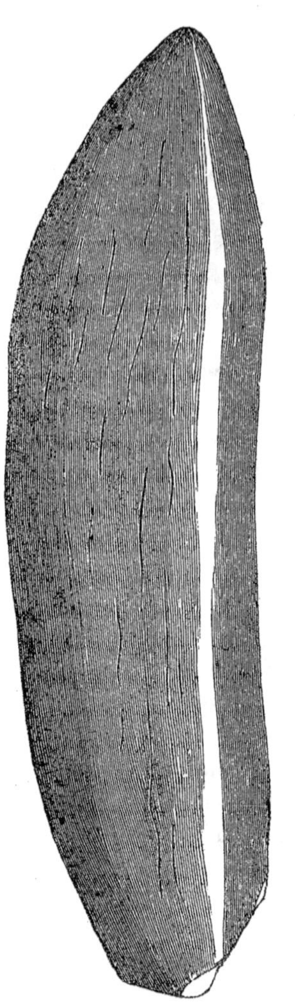

The oldest specimen of fossil belonging to the whale or cetacean family, belongs to the genus Physeter, and is regarded as the P. antiquus, (fig. 34.) It occurs in the eocene of Craven county. The size of the teeth prove that they belonged to the largest of the class. The largest tooth measures six inches in circumference, and is five and a half inches long, though a portion has been broken from the base. Its form is quadrangular, and presents a curve in front, but is rather straight behind. It shows no conical cavity, but is solid throughout. It shows a tendency to exfoliate concentrically. Many fragments more or less rolled and otherwise defaced, have been seen in the miocene beds upon the Tar River.—It is probable they may have been removed from a lower to an upper formation.

CHAPTER XVI.

REPTILIA.

Description of Reptilian remains of the marl beds of North-Carolina.—Reptiles of the Green sand.

I was fortunate in discovering a vertebra of a large size on the lower Cape Fear, which, at the time, I supposed to be new. As the discovery was confined to this single piece of the skeleton, I deemed it insufficient to draw from it special conclusions respecting the family of saurians to which it belonged.

Since this discovery, Prof. H. D. Rodgers has presented to Prof. Owen, of London, a collection of vertebrae from the green sand of New Jersey, among which I find the saurian described, to which my North-Carolina fossil must belong.

Figure 34 (A.) represents the vertebra from the upper part of the green sand of North-Carolina. It belongs to the lumber region. Its type is procelian, that is, it is concave before and convex behind, like the crocodiles of the present day. The body is long, and from the anterior half it sends off strong processes at nearly right angles, which are thin and strong. The articulating extremities are less concave and convex than those of the alligators of the Southern States. In this character I find it agrees essentially with those of New Jersey.

The abdominal face is smooth, and marked by two, or a pair of elongated holes, situated rather nearer the concave than the convex end. The body is cylindrical, especially posteriorly. Prof. Owen refers the New Jersey saurian to the lizards and to the mososaurian type. The name which has been conferred upon this remarkable saurian is *Macrosaurus*. If my determination is right with respect to the identity of the New Jersey and North-Carolina specimens, it will be known by the same name. This vertebra is three and three quarter inches long, including convexity, which equals half an inch, and six inches from the end of one parapophysis to the other; across the concave articulation nearly two and a half inches; across the convex, two inches; length of the lateral process, nearly two inches.

FIG. 34 (A.)

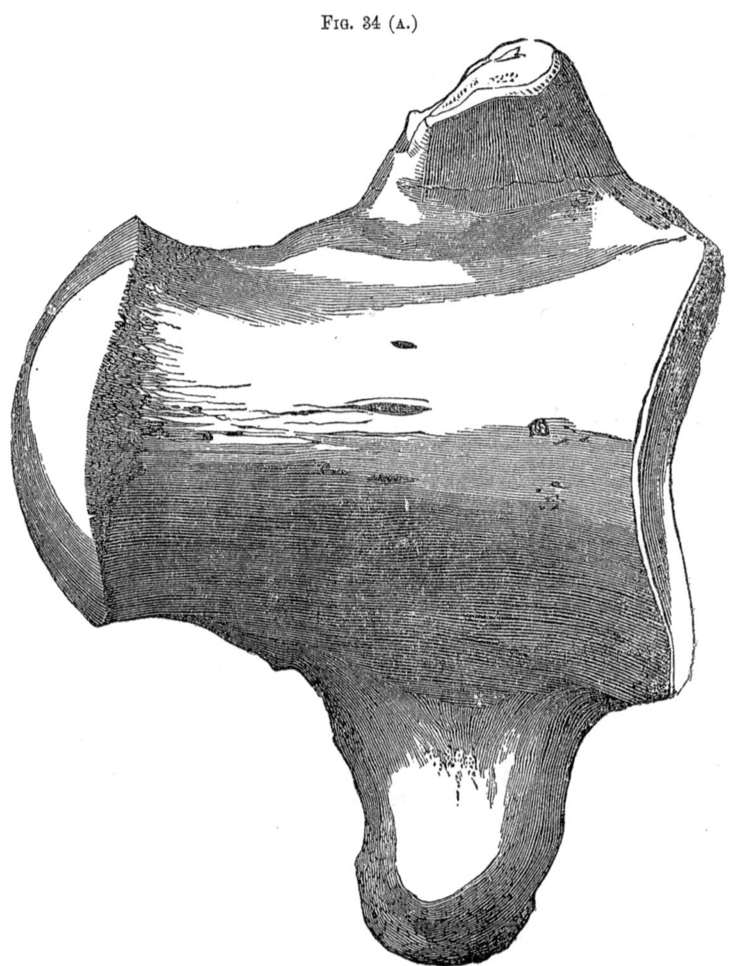

The entire length of this saurian cannot have been less than twenty-five feet, and it is a fact worthy of notice, that saurians of this description inhabited a region as far north as New-York, while at the present day their limits are confined to the central parts of North-Carolina. This fact, no doubt, indicates a milder climate in New-York and New-Jersey than is known at the present day. All the large land reptiles are confined to the warmer regions of the globe.

CROCODILUS ANTIQUUS.—LEIDY.

Another extinct saurian (fig. 35, A.,) is indicated in the discovery of vertebræ, which belong to, or are found in, the miocene marls. The most perfect one which I have obtained, is the 2d caudal, which as it is possible to identify it, may be compared with the Alligator luscius, the common large reptile of the Southern States, inasmuch, too as it belongs to the same type of vertebræ.

Fig. 35. (A.)

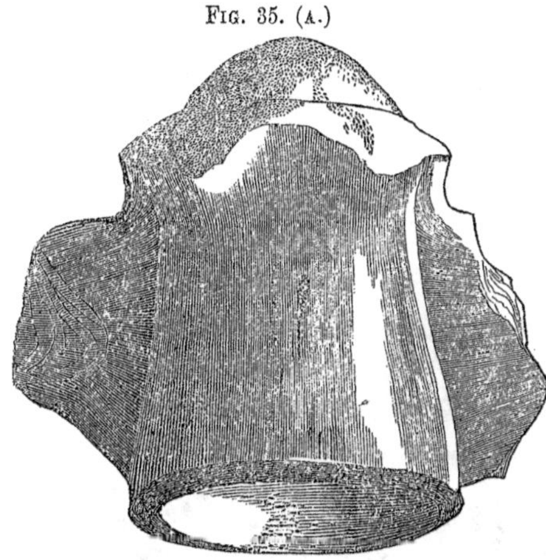

This vertebra differs from the corresponding one to which I have referred it; it is rather larger and thicker, and the proportion of its parts differ also. Its length is one quarter of an inch greater, but its diameter at the concave end is three-eighths greater, and the size or diameter of the body is still greater. The fossil is thick through its whole length, and varies but little at the ends; or it is much less compressed laterally than the vertebra of the living Alligator, and what is equally worthy of note, is, that the transverse processes come out more immediately from the body of the vertebra than the other. One more point may be made; a ridge of bone begins near the middle at the concave end, and runs a little downwards, until it reaches a slightly constricted part just before the border which surrounds the convex extremity; this gives the appearance of breadth to the bone when we look upon the abdominal face. There is a slender sharp ridge occupying the same relative position in the Alligator, but it seems to originate at the convex extremity. A slight groove runs longitudinally upon this face. Length, one and eight-tenths; width, over the concave end, one and five-tenths inches.

From all that I have been able to glean from the discoveries of others in this country, these vertebra appear to belong to a species which has been discovered in the miocene marls of New Jersey and Virginia. The species is now extinct.*

The cranial plates, one of which is illustrated by figure 36, belongs to a large unknown saurian. These were taken from

Fig. 36.

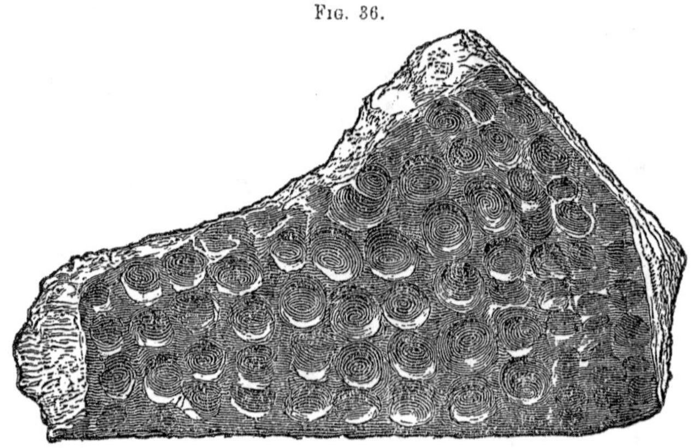

the miocene upon the Neuse, fifteen miles below Goldsboro'. They are over half an inch thick, and ornamented with deep sculpturings, and from their massiveness might be referred to the *Macrosaurus*. But this reptile belongs to an older formation. I have, however a laniary tooth of the proper dimensions to correspond in size with the saurian, which may have been provided with this impenetrable armour, and also the middle

* Proceeding of the Academy of Nat. Sciences, Phil., Vol. V, p. 307.

part of a femur to match both the plates and tooth, and all from the miocene or shell marl. The materials, however, for drawing up a proper description of this saurian, do not exist at present.

MOSSOSAURUS.

Tooth sharp pointed, pyramidal and curved backwards; enamel moderately and finely wrinkled; surface divided into two unequal parts by well defined and finely serrate carinae, the anterior of which is considerably curved on the last half inch, which forms the apex. Base of the outer surface smooth, and forming the segment of a large circle; this smooth band is usually covered with a thin enamel, and is a little over a line wide. The rest of the outer surface is divided by three ridges, the middle is strong, and extends to the point; the anterior dies out about half an inch from the apex; the posterior is inconsiderable, and extends a little more than half way to the apex; these ridges divide the surface towards the base into three slightly concave surfaces. The inferior has eight distinct ridges none of which reach the apex; these divide this strongly convex face into nine slightly concave facets, of which those adjacent to the carinae are the widest, (Fig. 36, A.) side view, natural size, (Fig. 37,) viewed from the point, showing the division into parts and its polygonal form.

Fig. 36. (A.)

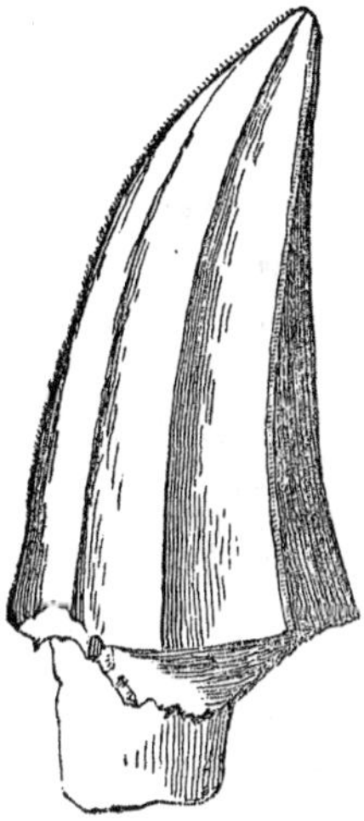

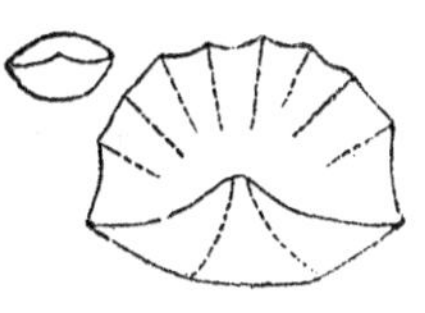

Fig. 37.

It is possible this tooth may differ from others which have been described. It differs from the one described by Dr. DeKay* in being finely rugose, and distinctly serrate, neither does he speak of angularities, though they are faintly indi-

* Annals of the Lyceum of N. Y., vol. 3, p. 136.

cated as existing upon the outer face in his transverse section, but that those faces are concave has not been stated by any writer.

The transverse section of the tooth, Mossosaurus Hoffmani, given by Prof. Owen, has no angularities at all on either face—the figure of the M. Maximiliani exhibits them upon the anterior face, but none upon the inner.

The tooth which I have just described is perfect, and not worn; the figures are good illustrations of its characters, and it appears, therefore, that the characters are either not uniform or else there are two species belonging to the green sand. It is evident that the tooth in question belongs to the species, Maximiliani, rather than the Hoffmani or gracilis.

POLYGONODON RECTUS,—LEIDY. MOSSOSAURUS RECTUS.

Fig. 37. (A.)

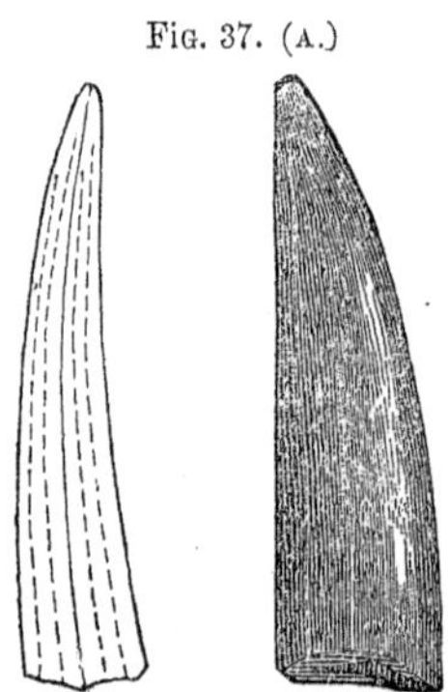

Tooth long, pointed, compressed; nearly equally divided on the outer and inner faces; the faces are formed by five equal and similar planes, bounded by angular ridges, only two of which, on each face, can be said to approach the apex; these are the two anterior and two posterior ridges curved backwards; bicarinate; but the posterior edge is nearly straight, while it has a convexity before which gives an apparent curvature which does not exist; edges smooth; enamel is probably thin or removed, leaving a dense dentine, with fine longitudinal cracks which appear at first like fine striae. The tooth is broken at the base of the crown, showing a shallow pulp cavity.

This tooth differs from any of the preceding in its form and surface. It is particularly noticeable, that the part near the base is distinctly angular, and is divided into ten nearly equal planes, and is bounded by well defined angles. All these angles extend a little above the middle of the tooth. It differs from either of the three species of Mossosausus in its proportions. It also differs from the teeth of the Leiodon, by being much more compressed. The teeth of the Polyptichodon are circular, and the teeth also of the Pliogonodon, which I found upon the Cape Fear, are also quite circular and conical. It is possible it may be a palatine tooth of the M. Maximiliani. It differs, however, in form from those teeth. It appears to have had that kind of attachment to the jaw, which has been called acrodont. Length, one and three-quarter inches; width, at base, seven-sixteenths.

POLYPTYCHODON—OWEN. POLYPTOCHODON RUGOSUS.—E.

The teeth (Figs. 38 and 39) which are represented in the margin were discovered in a bed of miocene marl at Elizabethtown, Bladen county, in 1852–'3. They were regarded at the time as having belonged to an extinct undescribed species. I have had hopes that other parts of this saurian would be discovered which would throw some light upon its organization and form, but as yet no bones which can be referred to the genus, or species to which the teeth belonged, have come to light. Saurian bones of a large size are not wanting which may have belonged to the teeth under consideration, but more than one species have been discovered. In one instance the middle of a large femur; in others cranial plates, the sculpturing of which are quite different, are among the bones which have been discovered. These, however, are disconnected fragments, and hence are insufficient to settle the question of ownership. The epoch to which the bones referred to belong is not at all established. Large saurian vertebra have been found in the green sand, and teeth resembling those found at Elizabethtown in the eocene marl upon the Neuse. Hence it is probable that the

FIG. 38.

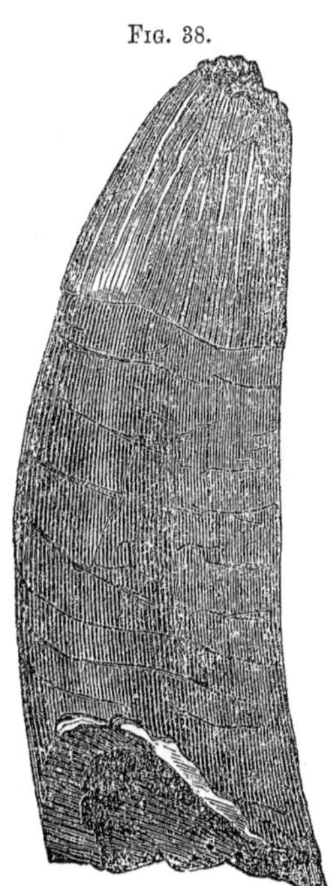

epoch of these reptiles is earlier than that of the miocene beds. They are found in those beds for the same reason that the exogyra costata of the green sand is also found in the miocene. While it is clear enough that fossils have been washed out of the green sand into the miocene. I have no evidence that they have been transported into the eocene, the next series above. The deposits seem to have quietly succeeded the green sand; but when the miocene period arrived, there was a breaking up of the older series, and their contents carried immediately up to this period, and under favorable circumstances fossils of both periods were intermingled together, and hence I regard the animals under consideration to have lived before the miocene beds were deposited.

Fig. 39.

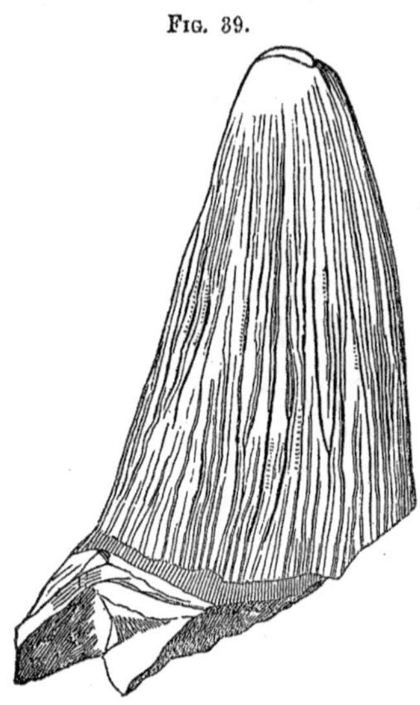

The teeth which I have figured I have referred to a genus of crocodilian reptiles established by Prof. Owen, and which, in England, belonged to the chalk or cretaceous system.

The following description is drawn from the teeth before me: Teeth thick and conical, and slightly curved; transverse section circular or round; enamel traversed longitudinally by numerous transversely rugose cracks, the strongest of which reach the apex; no trenchant edges or carinae proper.

The teeth are only gently curved; they are very strong and robust, and the enamel is traversed by rather irregular rugose ridges, which appear like cracks. The inside ridge is stronger than the others, and are formed of two confluent ones, and takes the place of a carina, and extends to the point in the young tooth; but in old and worn teeth most of the ridges terminate considerably below the apex. The surface of the young tooth (Fig. 39) is very rough, and the edges of the rugosities are realy, irregularly serrate, and run into each other. The section is round at all places, from the base to the apex. Its crown is hollow, and its pulp cavity presents a conical hollow which extends about one-third of the length of the crown. On exposure to the weather, the crown exfoliates in conical layers. Below the crown, that part known as the root is hollow, but has a thick strong shell, which on the concave side has three wide shallow furrows; the middle one is exactly in the concavity; they occupy about one-third of the cylinder; the remainder is perfectly circular.

Prof. Owen's description of the polyptychodon* is as follows: "Teeth thick and conical; transverse section of the crown circular, without larger or trenchant ridges; enamel ridged longitudinally, but only a few reaching the apex. The crowns, when weathered, exfoliate in a conical manner by detached layers, like *a cone in cone;* base having a conical pulp cavity which opens into the crown in distinct sockets."

The foregoing description of Prof. Owen, of the genus Polyptychodon, applies so well to our teeth, that there can remain scarcely a doubt as to their generic identity. It is, however, only a generic similarity; the species appears to be quite different from both of the species described by Prof. Owen, and from its remarkable rugose enamel, I propose as its specific name, *rugosus.*

It differs from the Alligator in the absence of a deep constriction at the base of the crown, from the Pliogonodon of Leidy, by its robustness and rugosities, and from the Elliptonodon, by its circular section, this having a circular section only at the base of the crown, while in the former the crown has a circular section from base to apex.

Sculptured Cranial Plate, (Fig. 40.)—These plates are separated from each in the line of suture, and are generally broken. They are thick and strong, and were no doubt sufficiently

* Palaeontographical Society's transrction, p. 46, vol. for 1851. (Description of the P. interuptus and continuus.)

so to resist the entrance of a músket ball. The same remarks as it regards ownership have already been made, respecting other bones of this class, so common in these deposits. That there were two, at least, powerful reptiles, is evident from their bones and teeth, but in no instance have two been found attached, and in such relations that it would be safe to affirm that they belonged to the same individual.

Fig. 40.

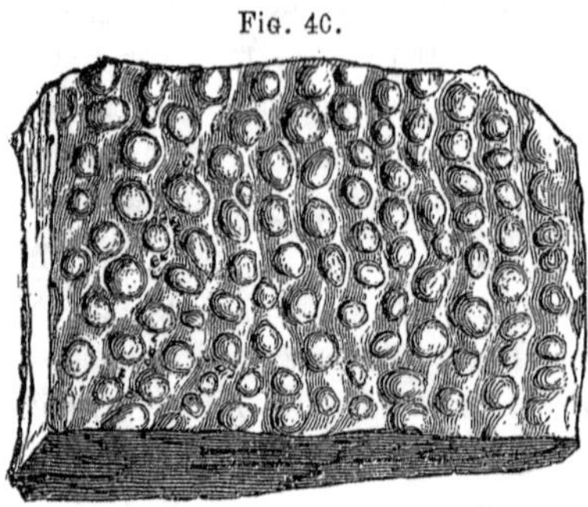

ELLIPTONODON COMPRESSUS.—EMMONS.

Figs. 41 & 42.

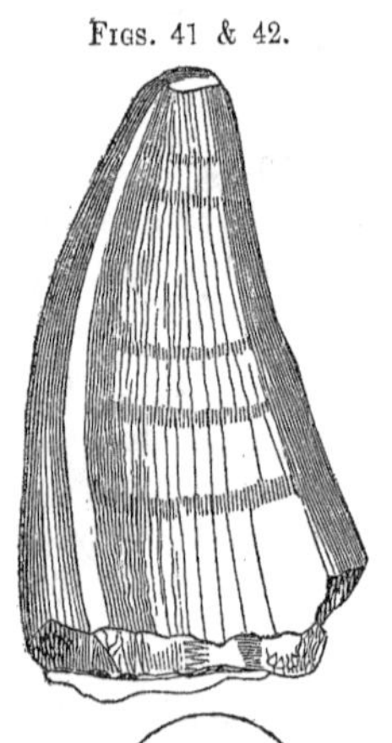

Tooth curved, robust, sub-conical and pointed; crown circular at base, becoming elliptical, and finally sub-elliptical, or with the inside more flattened or less convex than the other; bicarinate; the anterior ridge becoming obsolete near the base of the crown, and without serratures or rugosities; enamel rather finely wrinkled longitudinally, or faintly rugose, and none of the rugosities extend to the apex; dentine is concentric; pulp cavity open, conical, carinate. Figures natural size. Figure 42, transverse section.

This tooth is broken at the base of the crown, and has lost a small part of its apex.

It differs very clearly from the Polyptychodon, Pliogonodon, Mossosaurus or Pleiosaurus. The clear and distinct marks of difference are shown in the figures of each referred to except the Pleiosaurus. This tooth was found in the miocene near the Cape

Fear River, in Bladen county. As the bones which have been found in these beds indicate the existence at a prior period of two large and formidable saurians, so the teeth confirm this view, and I have placed in this connexion a sculptured cranial plate, (fig. 40,) which may have belonged to this species.

Additional discoveries, however, are required before it is possible to determine to which of these plates the teeth respectively belong. All the bones which are found in the miocene beds, are broken, though they are mixed with perfect delicate shells. This fact proves that the bones were subjected to violence before they were imbedded in the miocene, and hence belong, probably, either to the eocene or green sand.

PLIOGONODON NOBILIS. LEIDY.—(Figs. 43 & 44.*)

In the collection of Prof. Emmons there are two, much mutilated teeth of a saurian discovered in a miocene deposit of Cape Fear, North-Carolina. These teeth, which have lost their fang and summit, are elongated conical, nearly straight or only slightly curved inwardly. Their section is circular with an inner pair of opposed carinae; and their surface is subdivided into numerous narrow planes and provided with a few vertical interrupted plicae, which are more numerous internally. The base of tho crown is conically hollow; the dentine is concentric; and the enamel is finely wrinkled.

FIGS. 43 & 44.

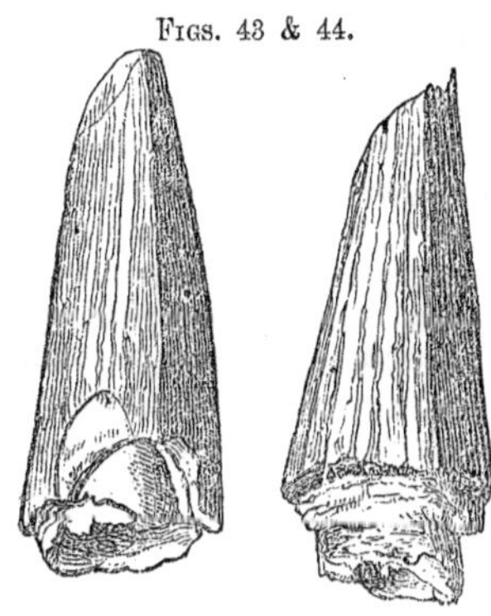

The specimens measure three-fourths of an inch in diameter at base, and are about one and a half inches long, but when perfect their crown has been a half inch longer.

* These teeth appear to differ, one has a coarser aspect, and the striar are coarser, and it is more curved, and proportion differs. Description by Prof. Leidy.

From the teeth of *Mososaurus* those of *Pliogonodon* differ in their narrower proportion, their straightness, their circular transverse section, their relatively narrower planes, and in their possession of plicae. From the teeth of *Polyptychodon* they differ in the possession of dissimilar planes and carinae, and in their less degree of robustness; and from those of *Pleiosaurus* in the existence of divisional planes and the circular section.

DREPANODON IMPAR. LEIDY.—(Figs. 45 & 46.*)

This genus and species are proposed on the crown of a tooth resembling the corresponding portion of the inferior canine of a bear, except that it has but a single carina, and that on the concave border internally. The specimen was discovered by Prof. Emmons, at Elizabethtown, Cape Fear, North-Carolina. It is black in color, curved, conical, most convex externally, and is oval in transverse section. The base is hollowed conically, and the enamel is smooth. The length of the specimen is three-quarters of an inch; the antero posterior diameter of its base is seven lines, and its transverse diameter five lines.

Figs. 45 & 46

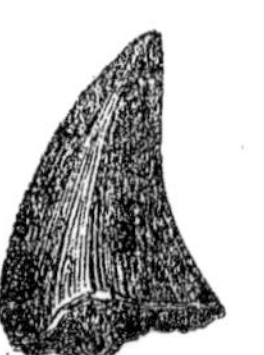

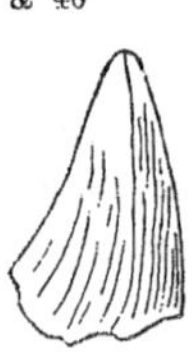

The tooth I suspect to have belonged to a crocodilian reptile, though it may possibly even prove to be a mammalian relic.

* Described by Prof. Leidy.

CHAPTER XVII.

PISCES.

Description of the remains of Fish in the North-Carolina Marl beds.

ISCHYRHIZA ANTIQUA.—LEIDY.

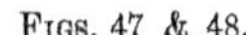

FIGS. 47 & 48.

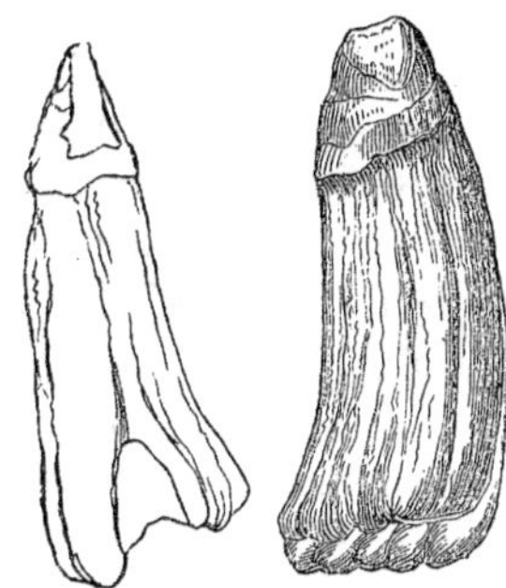

The curious genus Ischyrhiza, was first brought to my notice by the discovery of a tooth in the Green Sand of New Jersey, by Prof. Leidy. My collection contains several teeth discovered on the Neuse River. In most specimens the crown has lost its apex, but the fang is entire. In the perfect condition, the crown has been laterally compressed, conical and inverted with smooth, shining enamel. The fang expands from the crown in a pyramidal manner; is quadrilateral, curved backward, and divided at base antero-posteriorly; the division becoming deeper posteriorly. The larger specimen, in the figure, which is of a red color, when perfect, was nearly, or perhaps quite two inches in length. Its fang is an inch long, eight lines antero-perteriorly at base, and six lines transversely. The base of the crown is oval in section, and measures six lines antero-perteriorly, and five lines transversely.

The smaller specimen is black in color, and was about half an inch shorter than the other. Its fang is about ten lines long, and at base is about six lines square. It belongs to miocene of North-Carolina.

FOSSIL SQUALIDAE OF THE TERTIARY OF THE EASTERN COUNTIES.

The fossil squalidae, or sharks, are known only by their teeth, as these are the only parts which are usually preserved.

Their vertebrae are sometimes preserved, but they must be exceedingly rare in beds which are as loose as the clays and sands of the tertiary deposits. But the teeth, being protected by a very dense enamel, and having a firm strong core, resist change for ages; it is in these organs, therefore, that memorials of this highly interesting order of fish have been preserved.

The teeth being attached loosely to a cartilaginous jaw, are almost always separated and detached; and hence, they occur singly. Of the mode in which they are connected, we are informed by the living species which inhabit the adjacent seas. From this source of information, we may be assured that these single teeth were arranged in several rows in both jaws; that only a single one, those of the front, stood upright, while the remainder lay flat with the points directed backwards, or obliquely so. When the front teeth drop out, its place is supplied at once by the uprising of that one which is opposite the vacant space. The teeth, though very numerous, differ but little in form, though they differ more in size. The most remarkable difference may be observed on comparing the symphysal teeth, or middle row with those on each side. Thus, Fig. 49, shows a front section of the lower jaw of the *galeocerdo arcticus;* the outer row standing upright,

Fig. 49.

those behind lying flat, and the middle teeth consisting of a series of small ones. This figure, therefore, is a type by which the reader may compare the prevailing arrangements in the existing, as well as in this extinct family of fishes.

GENUS CARCHARODON.—SMITH.

Teeth very large, broad, triangular and rather uniformly dentated in both jaws. The enamel is usually cracked longitudinally; roots massive and divergent; inside nearly flat; surfaces smooth, and scarcely ever striated.

CARCHARODON MEGALODON.—AGASS. (Fig. 50.)

This species has the form of an equilateral triangle, though

FIG. 50.

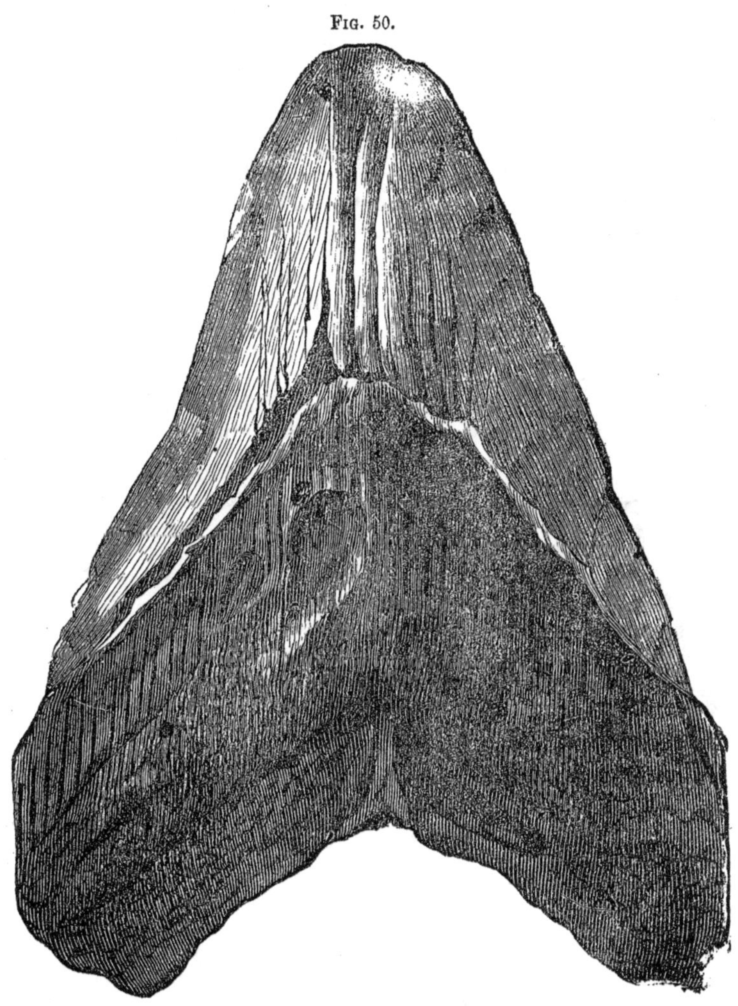

it admits of slight variations; teeth somewhat oblique, or inclined to the posterior end; upper, or outer side, nearly flat; under side prominently convex in the middle; enamel cracked longitudinally on both sides, particularly along the middle; serratures rather indistinct from the use of the tooth; core coarsely striated. It is usually found in the miocene beds, and is the most common upon the Cape Fear.

If the size of the teeth furnish an indication of the strength, size and ferocity of this species of shark, then it must have been one of the largest and most formidable animals of the ocean, combining, as Prof. Owen remarks, with the organization of the shark, its bold and insatiable character, they must have constituted the most terriffic and irresistable of the predaceous monsters of the ancient deep. The largest of the teeth measure sometimes six inches in length, and from four to five wide at base.

The jaws of the largest species of shark known in modern times measure about four feet around the upper, and three feet eight inches around the lower jaw. The length of the largest tooth is two inches, and the total length of the shark, when living, was thirty-seven feet. If the proportions of the extinct shark bore the same as those of the living, their length must have been over one hundred feet, equaling in this respect, the largest of the whales.

Fig. 51.

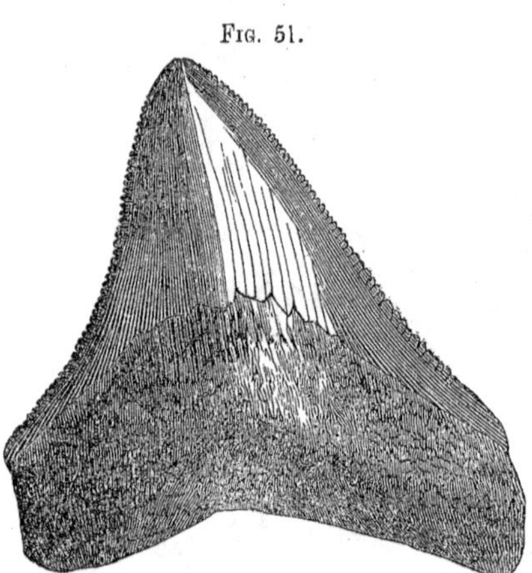

Figure 51 shows a smaller tooth of the carcharodon megalodon.

CHARCHARODON FEROX.—N. S. (Figs. 52, 53, 54.) Form nearly an equilateral triangle, thick; inner face very

FIGS. 52 & 53.

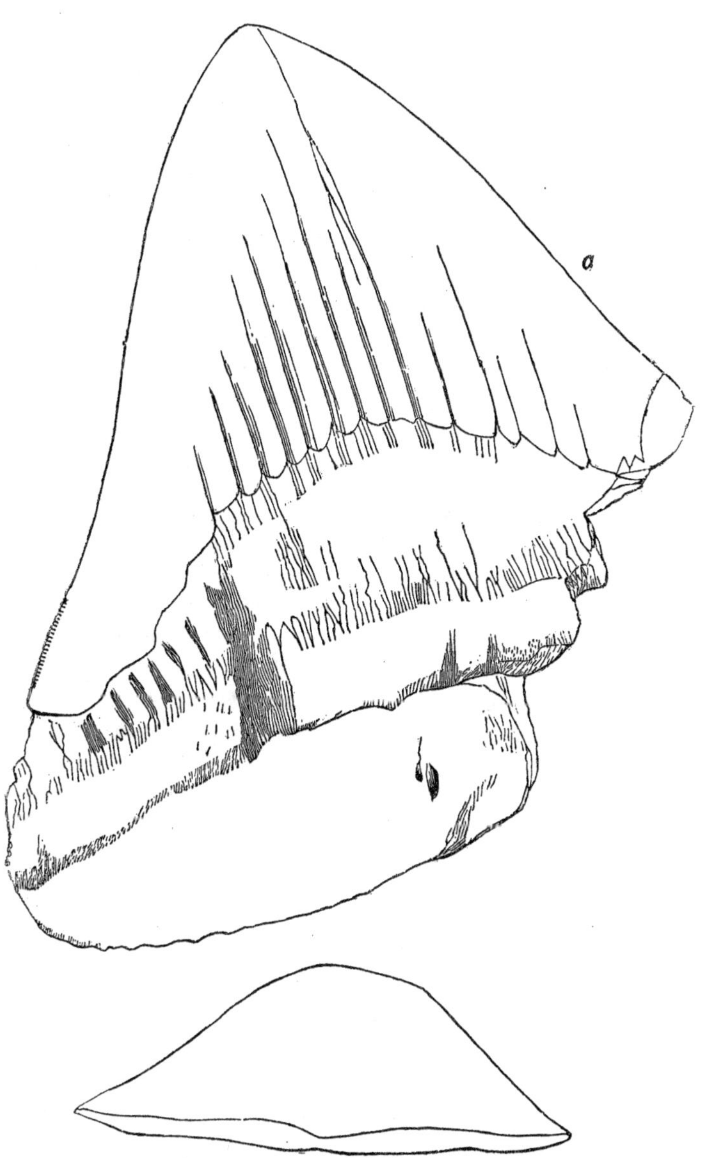

convex, outer nearly flat, and slightly champered towards the edges, and also slightly convex near the middle; seratures small, root thick, stout and straight across the base, and sloping on the inner face. The form of this tooth differs materially from the megalodon, especially in the relations of its height and breadth; height, four inches and a half, breadth at base, five inches; thickness of the root, one inch and a half, measured over the slope; length from the apex to the base of the root, five inches, measured along the edge; thickness at the middle, one inch. Found in the eocene of Craven county, N. C. The dimension of this species of shark equals that of the carcharodon megalodon. The tooth is thicker and stouter than this species, and more convex posteriorly, straighter across the base, and proportionally wider. Fig. 52 shows the outline of the tooth, fig. 54 is an edge view, and figure 53 a transverse section, showing convexity of the inferior face, and the flatness of the superior.

FIG. 54.

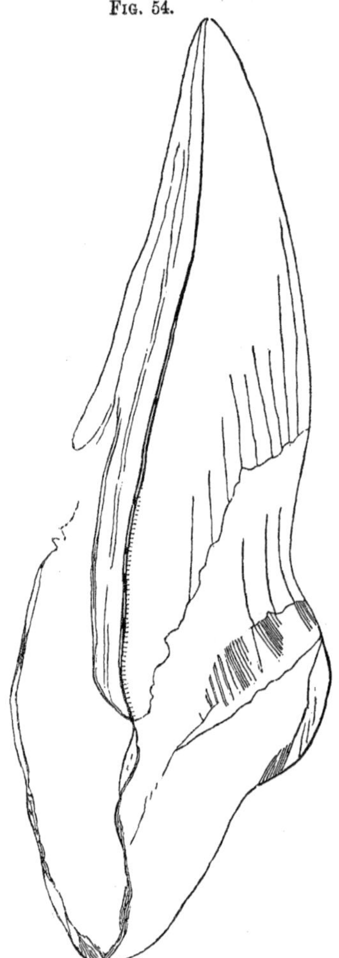

CARCHARODON SULCIDENS.—AGASS. (Figs. 55 & 56.)

Teeth large, thin and pointed; their forms correspond very closely to that of an isosceles triangle. They are flat on one side; the enamel extends to the root on both sides; it is more regularly sulcated upon the convex than upon the other side; fig. 55 young of the sulcidens.

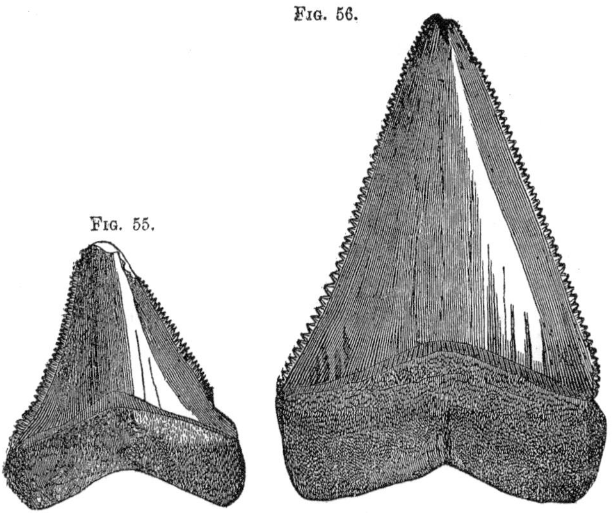

Fig. 55. Fig. 56.

CARCHARODON ANGUSTIDENS. (Figs. 57 & 58.

Fig. 57.

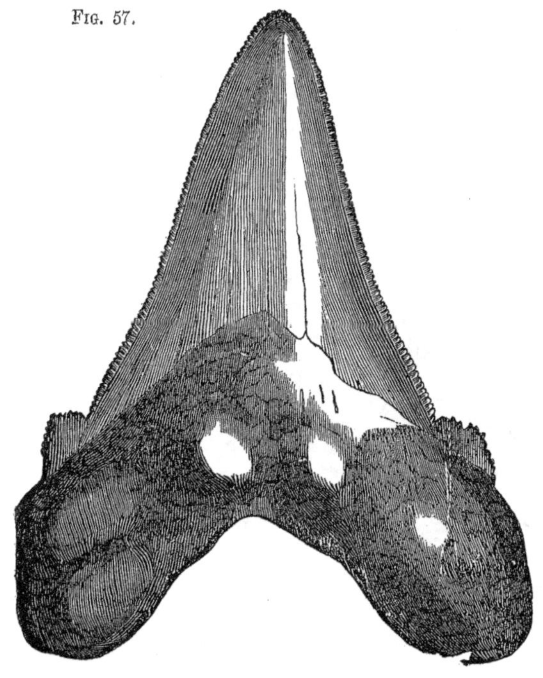

Crown only slightly oblique, rather thick, but comparatively narrow, but wide at base, and armed with serrated winglets, pointing upwards and outwards; the serratures are stronger than those upon the crown; roots massive, and separated by a distinct arch. Figure 58, a tooth which should probably referred to this species, though the arch of the root is flatter.

Prof. Gibbs, on the authority of Prof. Agassiz, has merged in the carcharodon angustidens, the following species: C. lanceolatus, C. heterodon, C. megalotis, C. semi-serratus, C. auriculatus, C. turgidus, C. semi-serratus, and C. toliapicus, on the ground that they are insufficiently characterized and not clearly distinguishable from each other.

CARCHARODON TRIANGULARIS, N. S. (Fig. 59.)

Crown of the tooth rather thin; the posterior faces of the crown meeting in the central line at an obtuse angle, but upon

FIG. 59.

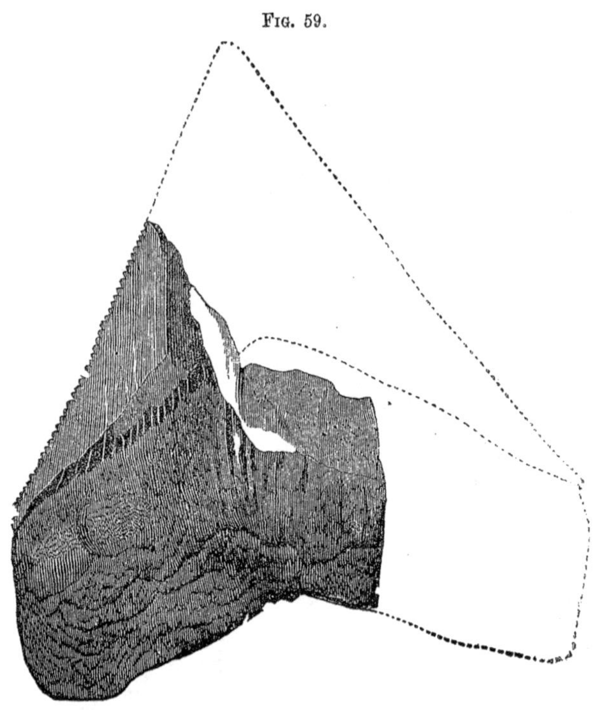

each side of this line they are quite flat; enamel thin, serratures small, root thick, striated and heavy, with a very low arch.

This tooth scarcely exhibits the usual convexities of either face; the faces being bounded by plane surfaces, the meeting of which give an obtuse angle when obtained by a central section through the crown. It belongs to the eocene, and was obtained from a bed near Newbern.

C. CRASIDENS, N. S. (Fig. 59, a.)

Tooth sub-conical, thick, slightly oblique; inner face very convex, outer flat at base, evenly but flatly convex near the apex, with an inconsiderable ridge extending from the base to a point near the apex, and somewhat ridged across the whole of the base of the outer face; serrae, sub-equal, and armed with serrate wings at base; root thick and prominent on the inside; enamel extends on the outer face to the root, and is extended continuously over the wings. This tooth belongs to the eocene at Wilmington. It is distinguishable from other teeth belonging to this order of fishes, by its very uniform degree of thickness from the base of the root, near its termination, at the apex.

Fig. 59, a.

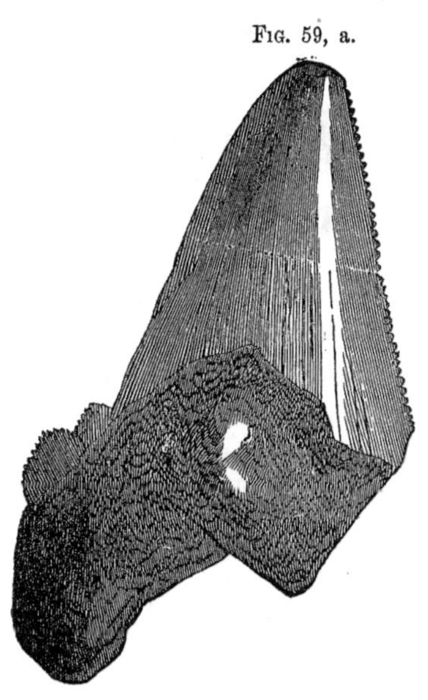

CARCHARODON CONTORTIDENS.—N. S.—(Fig. 60.)

Tooth an irregular cone, with the crown twisted near the summit; base of the root nearly plane, with the branches projecting upwards, rather than downwards, so much so as to stand upright when placed upon its base; inside the base projects enormously inward; enamel thin; serratures small, subequal; inner face very convex; outer nearly flat at base, but traversed longitudinally by an inconsiderable prominence.

Fig. 58. Fig. 60.

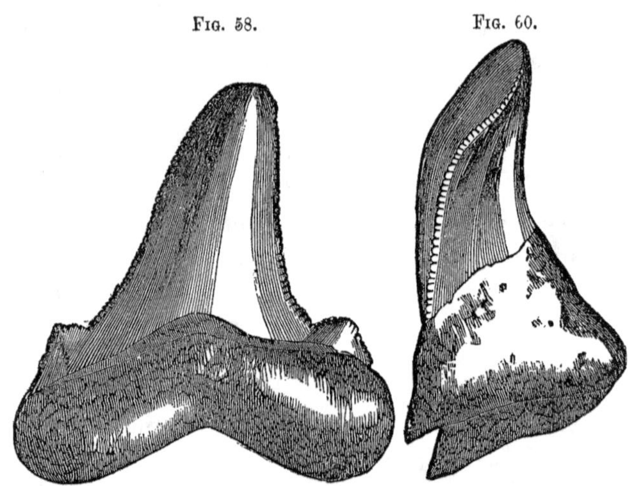

Only one tooth of this description has been obtained from the eocene at Wilmington. The form of the tooth is very peculiar, and may be readily distinguished by the great thickness of its root and projection inward. This projection is on a level with the branches of the root. The twist also, at the extremity, is also, a prominent feature in this tooth. It is probable, this tooth indicates the existence of a genus, which should be separated from the carcharodon, but the existence of a single tooth does not furnish all the characteristics which probably belong to it.

SPHENODUS RECTIDENS.—N. S.—(Figs. 61 & 62.)

FIGS. 61 & 62.

Tooth very long; comparatively slender; both faces convex; internal more so than the external; becoming narrower towards the edges; the base in some of the teeth trenchant, then nearly parallel two-thirds the length; enamel rather thick grooved on the inside, and cracked longitudinally on both, with a texture coarser than in the lamna; root unknown. Figure 62, transverse section. Green sand of North-Carolina.

GENUS HEMIPRISTIS.

Apex simple and smooth; margins of the tooth denticulated to a point near the apex.

HEMIPRISTIS SERRA.—AGASS.—(Fig. 63.)

FIG. 63.

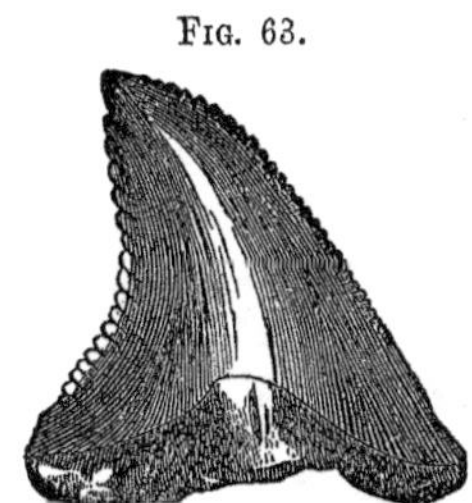

The H. serra is characterized by teeth which are serrated to a point near the apex, where the serratures cease, and the margins are left smooth.

HEMIPRISTIS CRENULATUS.—N. S.

Form similar to the H. serra; sides convex, long at base, and rather thick; enamel smooth, and marked with only a few cracks; edges at base faintly crenate; entire towards the apex.

GENUS OXYRHINA.

Tooth flat, broad, oblique, lanceolate and smooth, widening at base rapidly; root thin and nearly straight, and destitute of spreading branches or forks.

OXYRHINA XYPHODON.—AGASS.—(Fig. 64.)

Fig. 64.

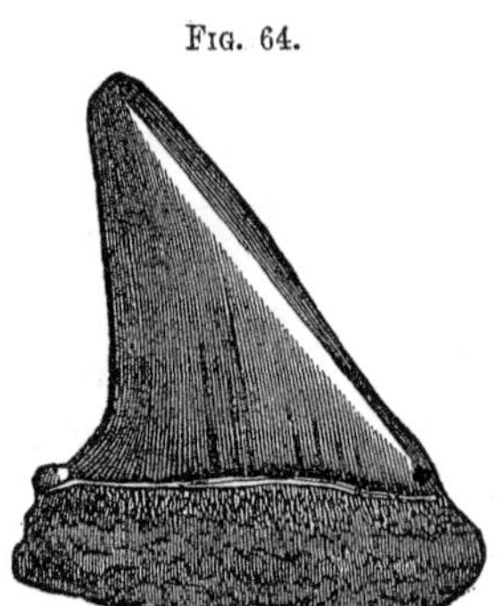

Lanceolate; base of the flat side marked with shallow furrows; enamel extends a little lower on the inner than outer side.

OXYRHINA HASTILIS.—AGASS.—(Figs. 65 & 66.)

Tooth rather elongated; lanceolate; nearly equilateral; bone of the enamel more arched than that of the oxyrhina

Fig. 65. Fig. 66.

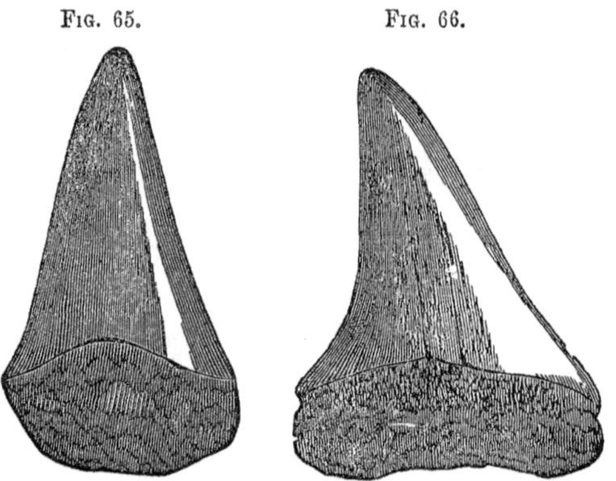

xyphodon, and the root seems to be less developed. It closely resembles the xyphodon.

OXYRHINA DESORII.—GIBBS.—(Fig. 67.)

Tooth thick and strong; roots well developed and forked; enamel similar in texture to the carcharodon, and also cracked longitudinally.

It differs from the former in the character of the enamel, curvatures, the absençe of serratures, and the form and development of its root.

Fig. 67.

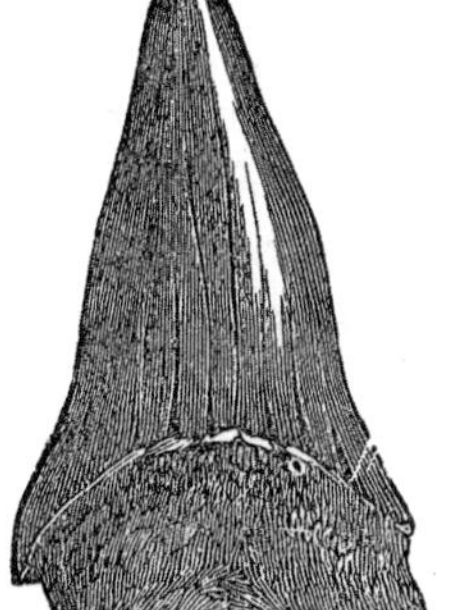

GENUS GALEOCERDO.—AGASS.

This genus is an inhabitant of our present seas, and the species arcticus (Fig. 49) very closely resembles the galeocerdo aduncus, whose teeth are abundant in the miocene marl beds of North-Carolina. In both jaws the teeth are similar and equal. They form five rows, which contain twenty-three teeth each, an odd small tooth occupying a middle position over the symphysis. The back teeth become small and are relatively shorter than the side teeth, presenting in this respect an approach to the form of the teeth described as the galeocerdo latidens. In two species of galeocerdo which differ in size, the serratures are constant and preserve a great uniformity; and the common character of the serratures seems to be, that which might be called compound, by which I mean, that each notch is itself notched, and it is possible that many of the species possessed this character more or less, but have lost it by wear in their usage.

Figure 49 shows the arrangement of the front teeth of the lower jaw in the galeocedo arcticus, and the position of a small series of teeth immediately above the simphysis.

GALEOCERDO ADUNCUS.—AGASS.

Tooth oblique angulated, and winged on one side, or with the sides unequal. Anteriŏr face convex, posterior rather flat. Serrate, serratures unequal, the first upon the wing the largest; upon the arched edge the serratures are largest upon the lower half of the crown.

GLLEOCERDO EGARTONI.

Tooth small, rather flat, lanceolate, slightly oblique, convex on both faces of the crown, but concave at the base on the outer face; root spreading widely, and obscurely wrinkled; serratures sub-equal, serrate or finely lobed; the enamel extends lower on the outer than the inner side. The latter character I am disposed to regard as its most distinguishable, for though the size of the teeth of this species may vary considerably, the character of the serratures will be preserved.

GALEOCERDO SUB-CRENATUS, N. S.

Tooth nearly upright, or with only a slight obliquity posteriorly; anterior edge formed by an arch belonging to the lower half, while the apical extremity or half the edge is straight, posterior edge is also straight for two-thirds the distance from the apex to the base, below which, the edge is drawn inwards; there is a constriction also on the opposite edge at the base of the crown; edges rather obsoletely crenate than serrate, smooth near the apex, and the smoother wing of the posterior edge stands at right angles to the axis of the crown; upper face rather flat, and marked by a faint rounded ridge extending from the base to the apex, and the surface slopes only from this ridge to the margins. The characteristics of this species will be gathered from the preceding description. The absence of distinct serratures, the form of the crown, its constriction at base, are the most important points, in which respects it differs from any which I have seen.

GALEOCERDO PRISTODONTUS.—AGASS.—(Fig. 68.)

Crown large, oblique; anterior edge irregularly arched, and extending much farther upon the base than the opposite edge; upon the flat, or nearly flat face, or outer one, the enamel extends below that on the convex side; seratures unuequal. Rare in North-Carolina, but I have several specimens, and from Dr. Gibbs's account of it, it seems to be still more rare in South-Carolina.

FIG. 68.

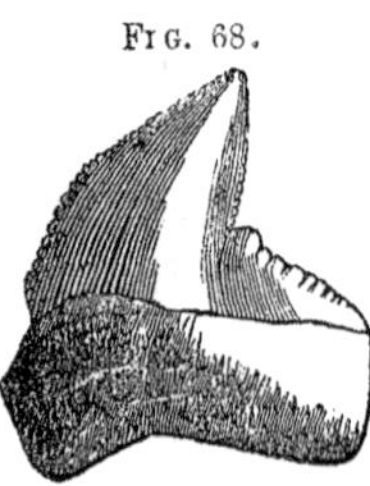

G. LATIDENS.—(Fig. 69.)

Fig. 69.

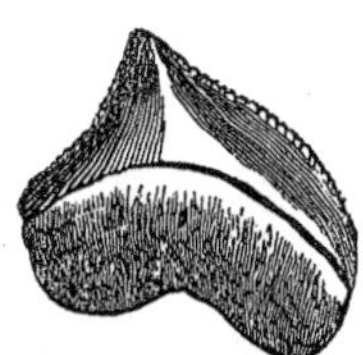

Differs from the preceding in its proportional length of base, being considerably greater.

The crown is low, and the enamel extends lower upon the outer face; the senatures subequal; apex pointed.

It is much more common than the G. pristodontus.

GENUS LAMNA.

Teeth rather flat, narrow and elongated; smooth, and usually furnished with appendages at base.

LAMNA ELEGANS.—AGASS.—(Figs. 70, 71 & 71 A.)

Tooth narrow, lanceolate; inner face quite convex, outer rather flat and smooth; the former regularly striate at base,

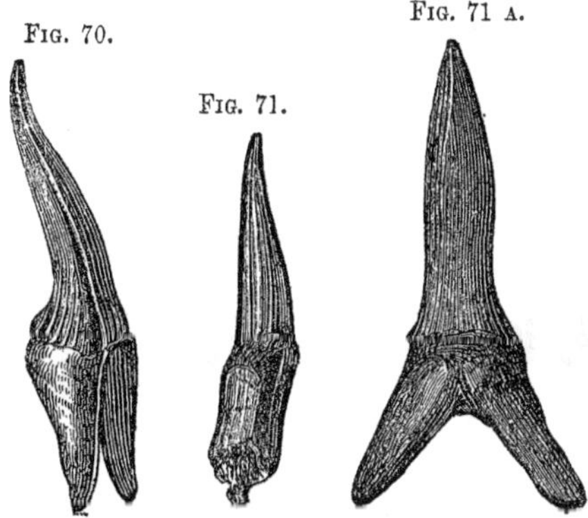

Fig. 70. Fig. 71. Fig. 71 A.

but towards the middle the striae degenerate into wrinkles; the outer ones are short, and but reach the edge of the tooth at base. The L. elegans is very common in the miocene beds of North-Carolina. Fig. 71 A, side view.

L. (ODONTASPIS) CONTORTIDENS.

Specimens which answer to the figures of this species, given by Prof. Gibbs, especially in the irregular form and absence of denticulations at base. In other characters there is only a slight difference between this and the L. elegans. They are found in the same beds.

L. COMPRESSA.

FIG. 72.

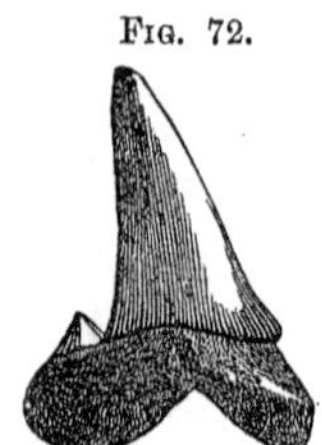

Compressed or flat, both faces convex and sub-equal, base irregularly denticulated; root wide and spreading. It differs widely from L. elegans and contortedens, but resembles the otodus; but Prof. Gibbs remarks that they are more lanciform, and the core more slender than the otodus.

FIG. 73.

FIG. 74.

Figures 73 and 74 appear to belong to the lamna. They are rather thick and stout, and resembles very closely an oxyrhina. Miocene.

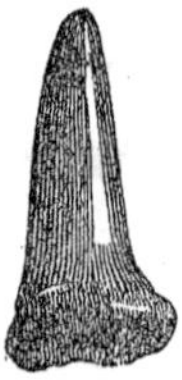

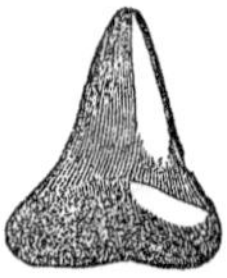

Figures 75, 76, 77, 78, 79, 80, and 81, belong to the eocene.

FIGS. 75 & 76.

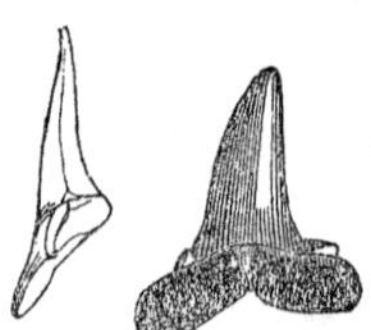

FIGS. 77 & 78.

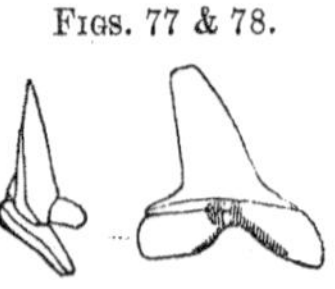

FIGS. 79 & 80.

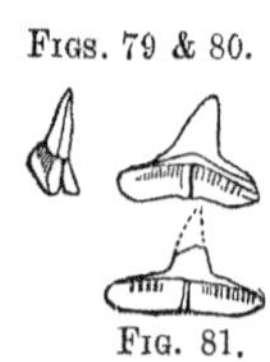

FIG. 81.

L. CRASIDENS.

Tooth thick and comparatively short; not very thick and projecting inwardly; inner face striate as in the preceding species.

GENUS OTODUS.

Tooth rather broad and flat, and armed with equal sharp denticles at base; root rather thick, projecting inward.

OTODUS APPENDICULATUS.—AGASS.

Tooth oblique, sharp or pointed, faces unequally convex;

denticles rather prominent and strong; line of base nearly horizontal; roots spreading widely, forming a very obtuse angle with each other.

I have referred also the following figures of teeth to the genus otodus: 82, 83, 84, 85, 86, 87, 88. They all belong to the eocene formation, and occur in a layer near the top. They are from the plantation of Mr. Wadsworth, of Craven county.

FIGS. 82 & 83. FIGS. 84 & 85. FIGS. 86 & 87. FIG. 88.

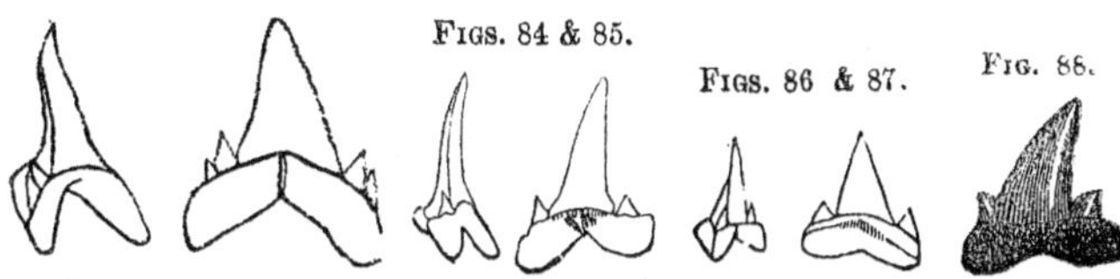

GENUS CORAX.

The following figures of teeth found in the eocene of Craven county. I am unable to refer them to species already described, viz: 82, 83, 84, 85.

FIG. 82a.

FIG. 83a.

FIG. 84a.

FIG. 85a.

GENUS ODONTASPIS.—(Figs. 86a, 87a, 88a, 89a.)

FIGS. 86a & 87a.

This genus should have followed lamna; I now introduce it for the purpose of referring to odontaspis, (figs. 86 and 87,) which appear to belong to this genus rather than lamna. So, also, figs. 88 and 89, which are destitute of basal denticles; but the cutting edge of the crown extends over the fangs and is slightly expanded on this part of the tooth; it preserves also its cutting edge. Eocene of Craven county.

FIGS. 88a & 89a.

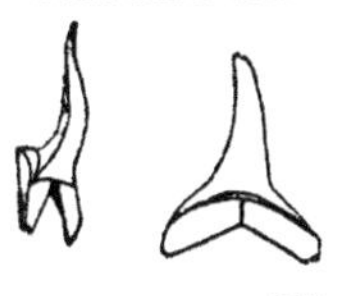

I have no facilities at hand which enable me to make a correct reference of the eocene odontolites, and have to trust to my memory in making the references to the genera to which they belong.

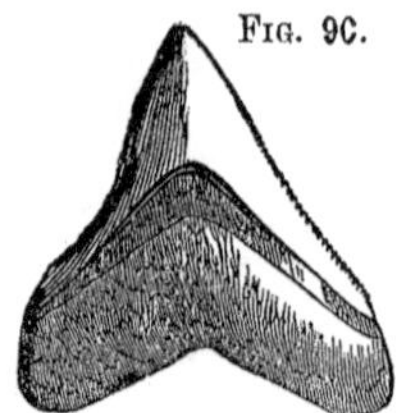
FIG. 9C.

CARCHARODON.—(Fig. 90.)

NOTE.—The annexed figure of a tooth, which may probably be referred to this genus, is confined to the eocene of Craven county. I have been unable to refer it to a species already made known.

SUB ORDER.—THE RAYS.

The rays are distinguished from sharks proper, by the flatness of their bodies. There are several species in the sea bordering the coast of North-Carolina, one of which is known by the name of *sting ray.* The rays form three families: 1, the pristides, familiarly known as the saw fishes, whose muzzles are elongated into a flat long extension, armed on each margin by pointed teeth; 2, rajides, or rays, whose muzzle is simple, but whose tails are not armed with a sting; 3, the mylliobatides, comprehending those rays whose tails are armed with a sting. The remains of the latter family are known in the tertiary and cretaceous of North-Carolina. Their teeth differ in form from those of the sharks, and would scarcely be regarded as teeth at all, were it not for their occurrence in the living species upon the coast. They are placed in the mouth in the form of a pavement, and occupy the areas within the mouth of both jaws. They differ in form from the pycnodonts in being angular. They are employed in crushing hard bodies, as the shells of the molusca. Their mouths are placed below, and well situated for seizing the animals upon which they feed.

FAMILY PRISTIDES.

Fish which have a prolonged, bony muzzle, armed with a plain horizontal series of teeth upon each margin.

GENUS PRISTIS.—(Fig. 93.)

Single teeth broken from the flat plate near its junction have been found in the superior layer of the eocene in Craven county. One margin is grooved the whole length, and straight, the other is curved and grooved only at base. Figure the natural size. I have also found smaller ones, which belong apparently to the same species.

FIG. 93.

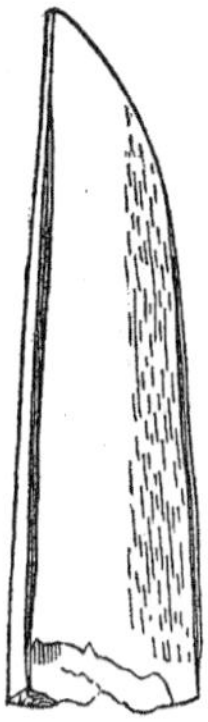

FAMILY MILIOBATIDES.

Rays whose tails are provided with serratine stings.

GENUS MILIOBATIS.

Sting dentated upon one margin. No stings of this kind have as yet been met with.

GENUS TRYGON.

Sting with both margins dentated.

TRYGON CAROLINENSIS.—N. S. (Figs. 91 & 92.)

Teeth in mosaic, the ends angular, they being bounded by six lateral planes.

FIG. 91. FIG. 92.

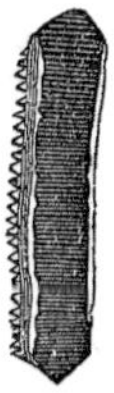

Sting serrate, (Figs. 94 & 95,) grooved longitudinally, rounded on one side. Fig. 95 shows the form of a tranverse section.

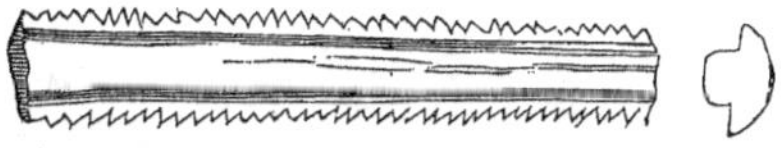

FIG. 94. FIG. 95.

These specimens were found in the upper part of the eocene marl in Craven county, and as the teeth and stings were found in proximity, it is inferred that they belonged to one specie.

CLASS GANOIDEA.—FAMILY PYCNODONTIDAE.

This family possess teeth of a cylindrical form, and which are arranged upon both planes of the jaws in the form of a pavement. The longer axis lies across the mouth from side to side, but set in rows arranged from before backwards. The middle rows contain the longest teeth, and they diminish in length towards the sides of the mouth. An idea of this arrangement may be obtained by an inspection of the mouth of the mylliobates, the common sting ray of the coast. In this fish the teeth are set also in pavement, but they are not angular. But the teeth in the Pycnodonts are not placed with so much regularity as in the Myliobatides.

FIG. 96.

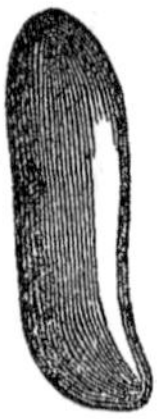

Fig. 96 is figure of a tooth belonging to the back part of one of the middle rows of the pavement, or mosaic. It may be called Pycnodus Carolinensis.

The teeth of this species of fish occur in the miocene associated with those belonging to the genera galeocerdo and lamna. The family of pycnodonts began their career in the Permian, but were the most numerous in the Jurassic period.

FIG. 97.

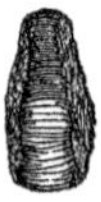

Another species of pycnodont is represented by its tooth in fig. 97, which appears to be much less common than the preceding.

SCALE OF A GANOID.—(Fig. 98.)

A single scale (fig. 98,) was found in the miocene upon the Cape Fear. The fish was closely related to the gar-pike, (lepidosteus,) of most of the American rivers. The scale occupied a position in the first row of scales back of the head. The fish of this class had already become rare at the commencement of this epoch. The gar-pike is the only surviving one of this family in the American waters.

FIG. 98.

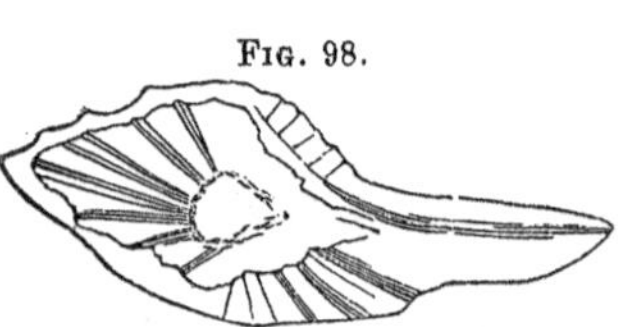

CLASS CYCLOIDEA.—(Figs. 99, 100.)

FIGS. 99 & 100.

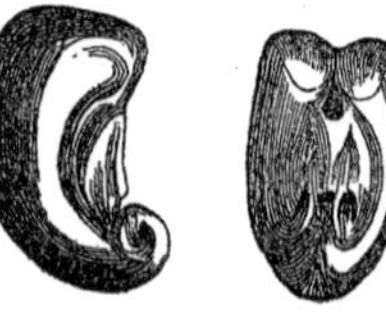

The annexed figures represent a peculiar form of fish teeth, which are quite common in some of the marl beds in Edgecombe county. They were attached by ligament, and probably occupied a position in the throat.

CHAPTER XVIII.

MOLLUSCA. CLASS—CEPHALOPODA.

This class embraces those mollusca, whose locomotive organs are attached to the head. They have the form of muscular arms or tentacles. Besides the arms surrounding the head, they have fins and an apparatus by which they can propel themselves through the water by its ejection in a stream.

Some are covered by a shell, coiled in a vertical plane, as the nautilus; others are naked or destitute of an external shell, but have an internal one, which varies much in form in the different families.

Their eyes are well developed and their mouths are provided with jaws somewhat similar to the mandibles of a bird. They are predatory and live on fish, crabs and shell fish.

The most remarkable part of the apparatus by which they seize their prey, are the circular discs arranged on the under side of their arms, by which they are enabled to produce instantaneously a vacuum when applied to the surface of a fish or a slightly yielding body. By this arrangement they are able to seize and hold most securely their captives, and devour them at leisure. As a means of escape from enemies more powerful than themselves, they are provided with a bag or sac filled with a dark fluid which they can eject at will, and thereby discolor the surrounding water and escape unseen.

This sac is called the ink-bag, and the liquid is employed for the manufacture of the India ink. Even the consolidated fluid in the fossil ink-bags is used for this purpose.

This class is a large one, and the species which compose it are found in all seas. They were also extremely numerous in ancient times, and their hard parts as external and internal shells are preserved as relics of extinct races. One of the most common fossils of the green sand is the Belemnite,* which is an internal shell, though its form is quite unlike one.

* From *belemnon*, a dart.

BELEMNITELLA AMERICANA.—(Fig. 101.)

The belemnitella is sub-cylindrical and tapering to a point from its base. The sides are marked by numerous ramose furrows, though they are arranged without much order, and being crowded they give the surface a granulated appearance. The base has a fissure which extends through the wall to a conical chamber. On the back, there is an elevated convex surface, narrow toward the base, but widens towards the apex, where it is lost.

This genus presents a great variety in form and size; but the foregoing characters are its constant characteristics. It occurs at Black Rock and Rocky Point, and is one of the characteristic fossils of the green sand. It is found also in the miocene beds, but is there by accident.

FIG. 101. FIG. 103. FIG. 102.

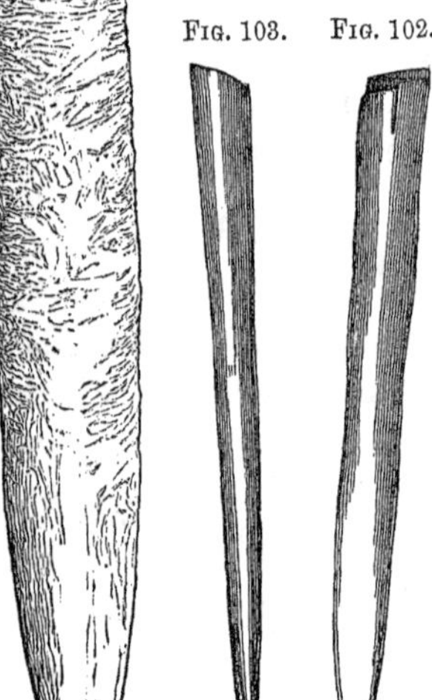

BELEMNITELLA COMHRESSA.—N. S. (Fig. 102.)

Shell slender, transverse section elliptical at base, and it becomes gradually more flattened to its apex; the fissure of the base is short; surface uneven and somewhat irregular. This species is entirely destitute of the granulations, or the convex surface of the preceding species.

The green sand of North-Carolina is poor in cephalopods. I have not yet observed either an ammonite or nautilus, though they occur sparingly in the eocene.

FIG. 105. FIG. 104.

In the eocene of Craven county I found numerous specimens of the bony or horny cores of the jaws of cephalopods. I have not been able to determine the family to which they belong. Fig. 104 represents their form and size. They occur only in the upper part of the formation associated with sharks' teeth, and teeth and stings of one or two species of ray.

CLASS GASTEROPODA.—FAMILY MURICIDAE.

The muricidae are generally readily distinguished by their roughness occasioned by the periodical expansion of its lip. These being permanent, the shell is strongly marked by the rough shelly expansions along the lines of growth, as in the murex. The shell preserves its spiral form; the outer lip is entire behind, and the front prolonged in a straight canal. The eyes of this family are sessile and seated on tentacles; the animal has a broad foot.

GENUS MUREX.—LINN. ROCK SHELL.

The shell is roughened, or winged with the periodical expansions of its lip, which are permanent after it has advanced to a mature state.

MUREX UMBRIFER—CON.—CERASTOMA UMBRIFER—TOUMEY AND HOLMES—FOSSILS SOUTH-CAROLINA FROM CON. MSS.—(Fig. 104a.)

FIG. 104a.

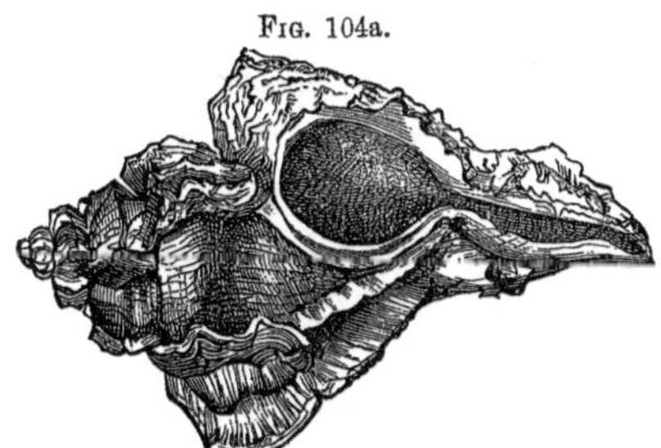

Shell fusiform; whirls subcarinated, or angulated and provided with six foliated and rather broad reflexed lamina, spirally arranged. Miocene Cape Fear River.

MUREX GLOBOSA.—(Fig. 105 A.)

Shell rather globose, or obtusely fusiform, and with four principal varices; intermediate ones irregular and spirally, traversed by many angular ridges, body whirl inflated, aperture oval, peristome continuous, and extended posteriorly on the body whirl, forming an angulated canal; outer lip ridged within and crenulated on the margin; collumela lip ridged, and one ridge at the posterior angle; beak reflexed. Miocene of the Cape Fear River; half the natural size.

Fig. 105 A.

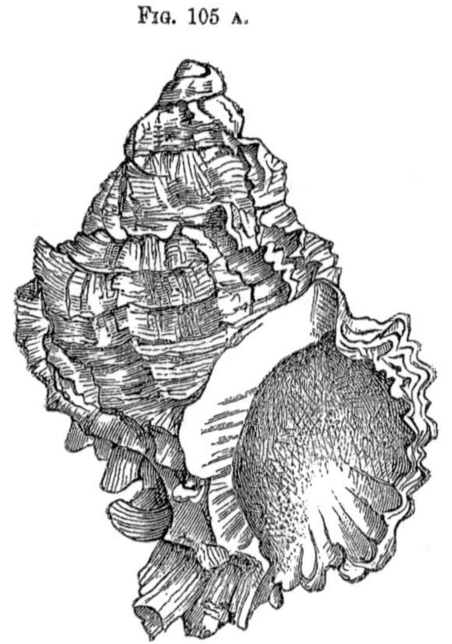

MUREX SEXCOSTATA.—(Fig. 106.)

Shell fusiform, with three spinous varices, and traversed spirally by angular ridges. Canal closed beak slightly reflexed. The body whirl has six ridges or ribs, with an intermediate lesser ridge. Shell imperfect.

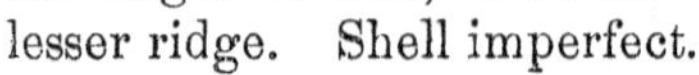

Fig. 106.

BUSICON CARICA, CON—PYRULA CARICA, GOULD, FULGUR CARICA, CON.

This shell is pyriform, swollen, thick and heavy. The outside is ornamented by transverse striae, and also with compressed tubercles, which stand upon the most prominent part of the body whirl. The outer lip is simple and sharp, pillar lip flexuous and concave above.

The suture of this species is not channelled, neither has it a turrited spire. It is one of the most common fossils of certain marl beds upon the Cape Fear river, but is less common upon the Neuse. It is one of the common living species upon the Atlantic coast.

BUSICON PERVERSUM, CON—PYRULA PERVERSA, REEVE.—(Fig. 107.)

FIG. 107.

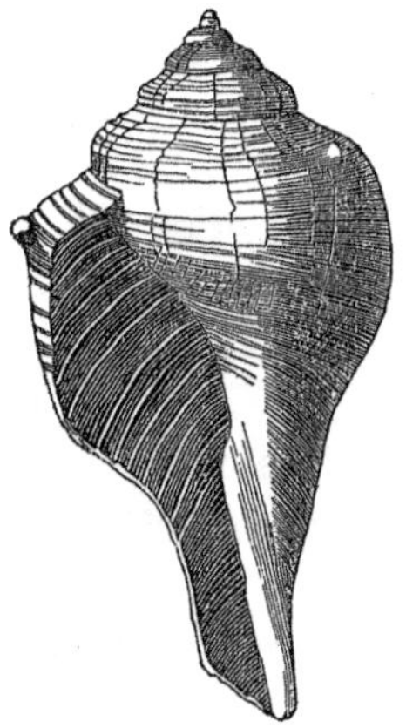

This shell is also pear-shaped and swollen. The prominent part of the whirl is ornamented with tubercles, and is also coronated; the whirls are turned to the left.

It is common upon the coast. It is very abundant in a post pliocene deposit at Beaufort, but is also often met with upon the Cape Fear.

BUSICON CANICULATUM, CON.—PYRULA CANICULATA, GOULD.—FULGUR CANICULATUM, CON.—(Fig. 108.)

Shell somewhat pear-shaped, spire depressed, and ornamented with revolving lines; body whirl swollen; canal long and straight; suture channelled. Common on the coast, and rather common, also, in the miocene.

FIG. 108.

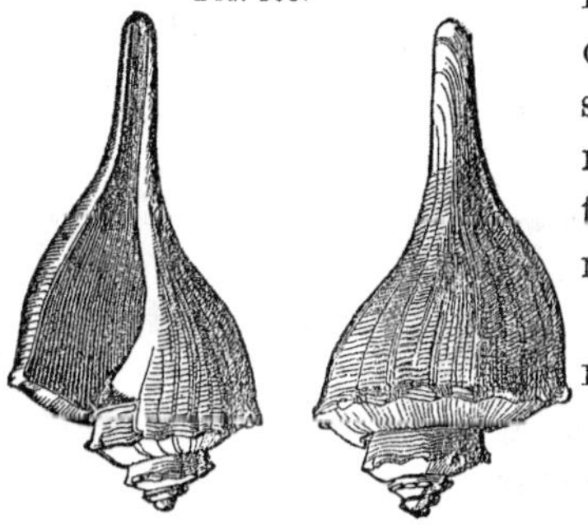

PYRULA CAROLINEUSIS—TUOMEY AND HOLMEL, H. TERTIARY FOSSILS OF SOUTH-CAROLINA.

Description: "Shell, pear-shaped; spire short, depressed; suture profoundly canaliculated, margined by the obtuse carina at the angle of the whirl; body whirl truncated above; angular whirls of the spire angulated in the middle, and in-

* Fossils of South-Carolina,—Tuomey and Holmes, p. 145-'6.

clined slightly to the summit, having fine revolving lines indistinct, but prominent and waved on the base of the body-whirl; canal long and tapering." Miocene marl, Cape Fear.

PYRULA SPIRATA, LAM.—FULGUR PYRULOIDES. SAY.—FULGUR PYRUM, CON.

Shell pyriform; spire depressed obtuse; whirls flattened, and traversed by numerous revolving lines; suture caniculated. It still lives upon the coast, and is common in the post pleiocene of North-Carolina.

FIG. 109.

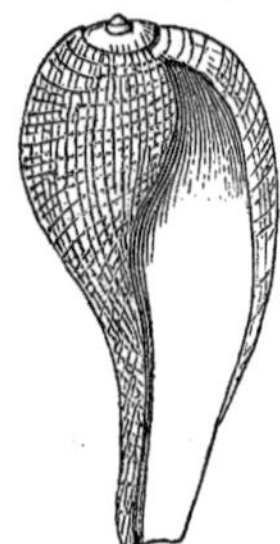

PYRULA RETICULATA—LAM—SYCOTYPUS RETICULATUS. (Fig. 109.)

Shell thin, cancellated; spire very short; surface marked by revolving lines, which are intersected by longitudinal ones, giving the shell its reticulated appearance or character. Occurs both in the miocene and post pleiocene beds, particularly at Beaufort. It is often much larger than the figure.

FIG. 110.

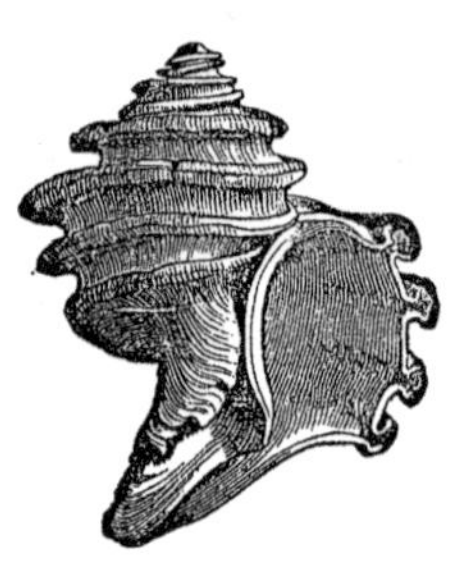

FUSUS LAM.

The genus Fusus is distinguished by its straight open canal and the absence of plaits upon the columella.

FUSUS QUADRICOSTATUS.—(Fig. 110.)

Shell thick, spire depressed, body whirl, inflated and ornamented by four elevated equidistant spiral belts, umbilicus large.—Newbern.

FUSUS EQUALIS.—N. S.—(Fig. 111.)

Shell thick, spire rather short, conical; whirls eight rounded and somewhat ventricose, and ornamented by numerous

Fig. 111.

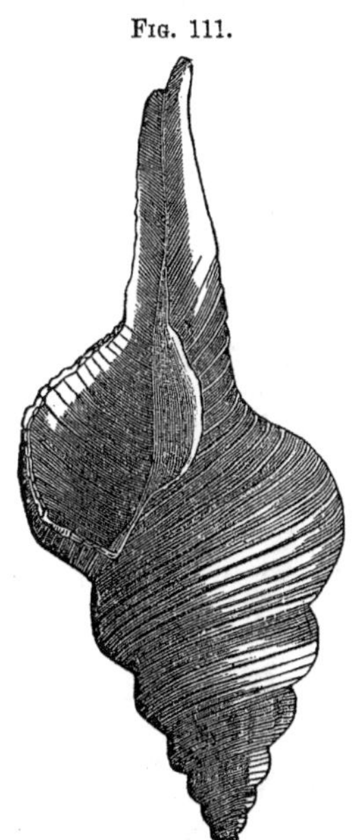

spiral subequal lines, coarser and more distant upon the back and rostrum; aperture and rostrum rather less than twice the length of the spire; outer lip ridged internally; pillar lip spirally ridged. Miocene of Cape Fear River.

FUSUS EXILIS.—(Fig. 111 A.)

Fig. 111 A.

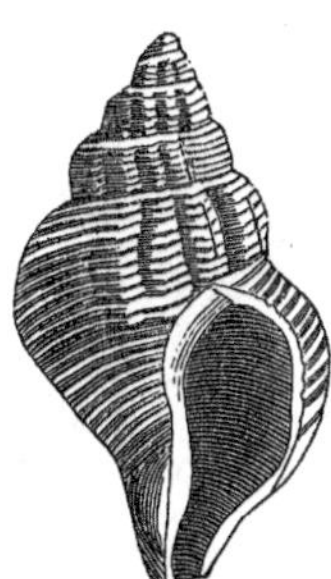

Shell fusiform, spire elongated, composed of seven whirls, ornamented by revolving striae and longitudinal ribs; aperture one half the length of the shell.

FUSUS LAMELLOSUS.—N. S.—(Fig 112.)

Fig. 112.

Shell small, fusiform; spire composed of five or six whirls, ornamented with ten strong scalariform ribs, each rib on the body is composed of three sharp crenulated plates, the one in the middle being the largest.

FUSUS MONILIFORMIS.—N. S.—(Fig. 123.)

Shell small; whirls four, ornamented with raised beaded spiral lines, between which there are lines nearly simple; spire rather shorter than the aperture; aperture oval; canal short; the two upper whirls are smooth. Miocene of Cape Fear. Rare.

FASCIOLARIA.

This genus is characterized by its elongated fusiform shape, its round or angular whirls, open canal, and its folds upon columellar lip, which is more or less tortuous. The folds upon the lip are quite oblique, and two or three in number.

FASCIOLARIA DISTANS.—LAM. (Fig. 113.)

Fig. 113.

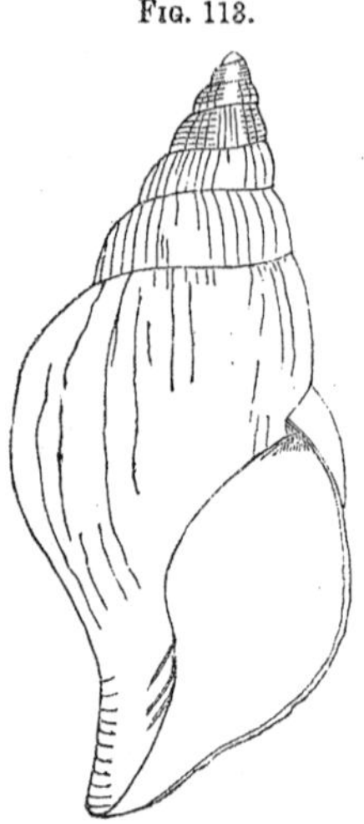

This shell at first sight appears smooth, but a careful inspection shows that it is finely striated longitudinally; its spire is composed of six or seven convex or prominent whirls, and its pillar has but one plait.

It is a common shell upon the coast, and in the post pleiocene at Beaufort, but not uncommon in the miocene of Cape Fear.

FASCIOLARIA ELEGANS.—N. S. (Fig. 114.)

Fig. 114.

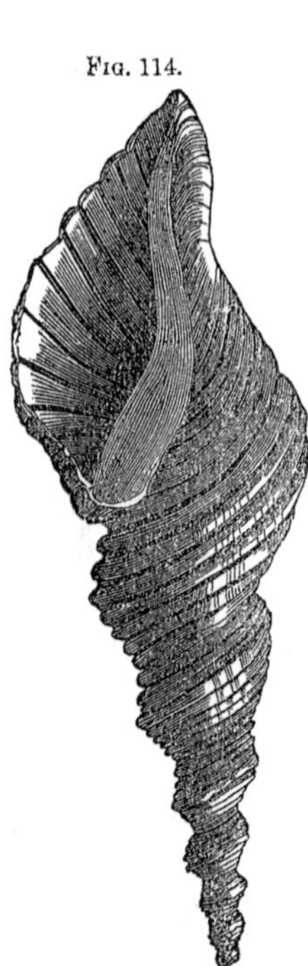

Shell elongated, acute; whirls eight rounded, ornamented with wide, and finely striated ribs; striae transverse to the ribs, or longitudinal; ribs of the body whirl, about fifteen, the middle of the body-whirl upon the outer lip, the four widest ribs alternate with three narrow ones; plaits three, concealed within the pillar lip; spire longer than the aperture.

This shell is rare in the miocene of North-Carolina. It would pass for fusus if the pillar lip was not examined just within the aperture, the plaits reaching only to its edge, but they are strong and well developed through its entire length.

It is possible this shell may have been previously described, but its broad, flat and very prominent ribs are so peculiar, that if observed and described, it could scarcely escape detection. Figure half the natural size.

FASCIOLARIA SPARROWI.—N. S. (Fig. 115.)

FIG. 115.

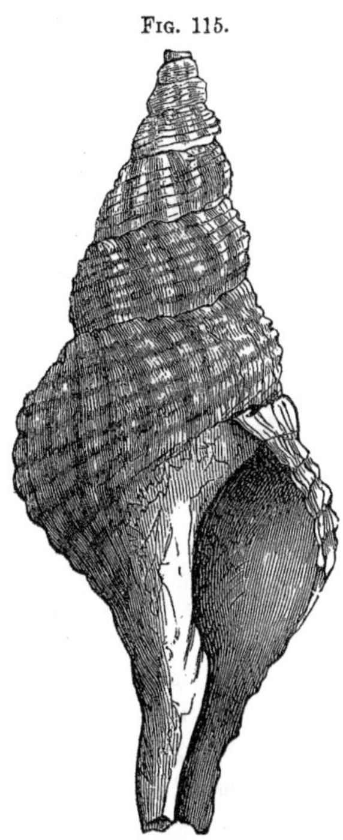

Shell rather thick, turbinate; whirls six or seven rounded, ornamented with spiral, and rather rounded ribs; ribs of the body-whirl, about ten, striated longitudinally, but obliquely striated on the upper part of the whirl; plaits, three upon the pillar lip; the ribs alternate, being coarser and finer for the ribs which belong strictly to the aperture; aperture larger than the spire.

This species is quite distinct from the former, the ribs are less numerous, flatter, and the striae are partly oblique and partly longitudinal, or in the direction of the axis of the shell. The five upper whirls have varices in both species. Rare in the miocene marl bed of Mrs. Purdys, Bladen county. One-half the size.

This fine fossil is dedicated to Thos. Sparrow, Esq., of Beaufort county.

FASCIOLARIA ALTERNATA.—N. S.

Shell rather small, but thick turbinate; whirls six or seven slightly inflated, body whirl elongated and ornamented with strong spiral lines, and with fine ones between, but which are frequently obsolete. All the whirls are tuberculated. Spire shorter than the aperture Plaits two.

FASCIOLARIA NODULOSA.—N. S. (Fig. 116.)

Shell rather thick, whirls about seven, all nodulose or

ornamented with varices and spiral subequal striae. Miocene of the Cape Fear river.

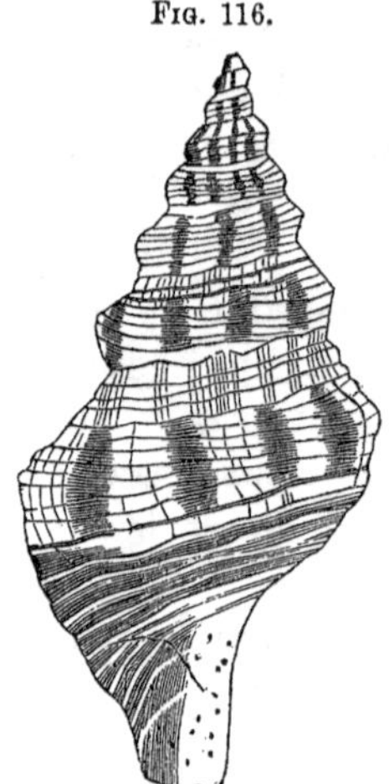
Fig. 116.

FASCIOLARIA ACUTA.—N. s. (Fig. 117.)

Shell elongated, acute, whirls about seven, ornamented by spiral subequal ribs, with obsolete ones between them, six upper whirls have also equal varices; longitudinal striae very fine, aperture shorter than the spire. Miocene of the Cape Fear river.

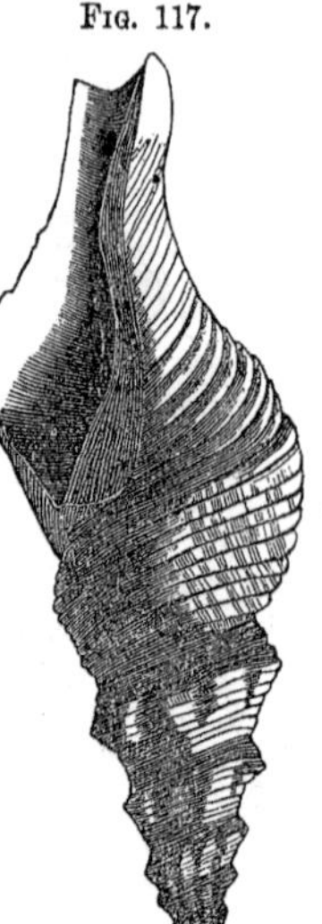
Fig. 117.

CANCELLARIA CAROLINENSIS.—N. S. (Fig. 118.)

Shell thick, angulated, whirls few, oblique, carinated and ornamented by two subspinous spiral bands, body whirl transversely, rugose towards the aperture, rugae subcrenulated, aperture triangular, and acute in front, umbillicus large, open, and funnel shaped.

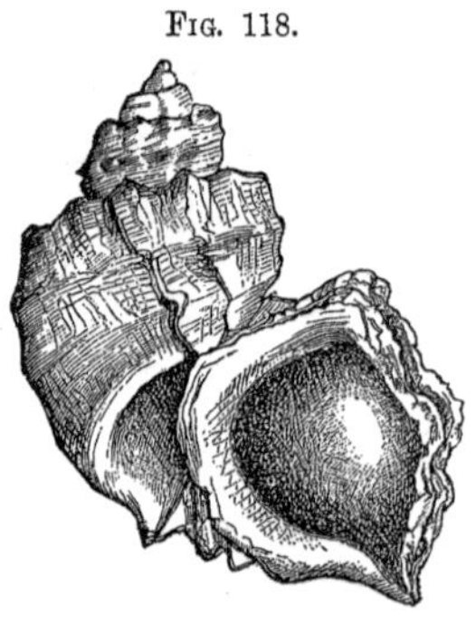
Fig. 118.

I should have hesitated to have placed this interesting fossil in the genus cancellaria were it not that a closely allied species, the C. acutangulata, Faujas, is thus referred by high authority. The C. acutangulata is one of the characteristic fossils of the miocene beds of Dax, south of France. Its surface is is ornamented like a cancellaria, but the aperture in both the Dax and North-Carolina specimens is triangular, but both have rather obsolete folds upon the pillar lip ; they are rather more obscure in our specimen than in that from Dax. The existence of this interesting fossil in North-Carolina proves the close analogy between the miocene of France and that of the southern States, and it seems that the new species really replaces the C. acutangulata in our miocene beds.

I am indebted to I. Lea, Esq., of Philadelphia, for specimens for comparison.

It occurs at Mr. Flowers' marl bed on the Cape Fear, Bladen county.

CANCELLARIA RETICULATA.—(Fig. 119.)

Shell thick, ovate, spire acute, whirls about six, and angulated and ornamented by prominent, longitudinal and revolving ridges, which produce a cancellated surface. Columulla with several strong oblique sharp folds; outer lip traversely ridged within.

FIG. 119.

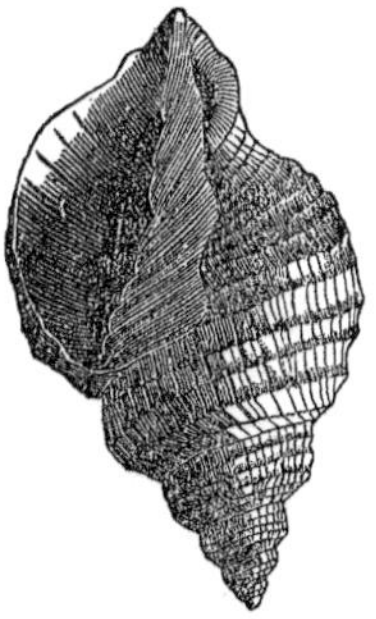

RANELLA CAUDATA.—(Fig 120.)

Shell turbinate, winged; whirls four or five, angulated and strongly ridged longitudinally; surface traversed by lesser revolving ridges. Two opposite ridges are produced more than the others, one of which forms the margin of the outer lip; canal long and straight. Common on the coast, and rather rare in the miocene of North-Carolina.

FIG. 120.

FIG. 121.

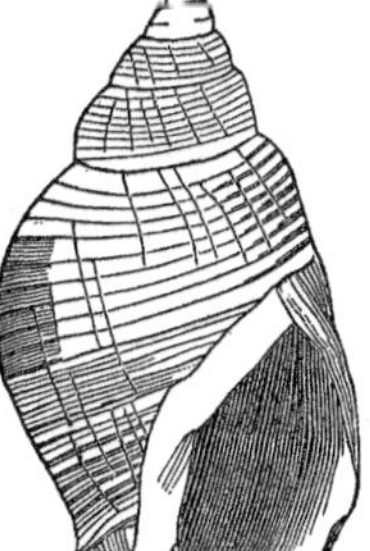

FAMILY BUCCINIDAE.—BUCCINUM MULTIRUGATUM.—CON. (Fig. 121.)

Shell thick, ovate; spire composed of five whirls, marked with deep impressed revolving lines; apex rather obtuse; columella, with a strong fold at base and a slight prominence at the base of the body whirl; bicarinate, aperture greater than half the length of the shell. Miocene of Cape Fear River.

BUCCINUM PORCINUM.—SAY.—(Fig. 122.)

Shell thick, fusiform; spire composed of five or six whirls, ribbed longitudinally, and marked with numerous raised revolving lines; beak short and only slightly reflexed; outer lip marked within by numerous ridges. Buccinum vibex, buccinum trivittatum and obsoletum are associated with the preceding species.

FIG. 122

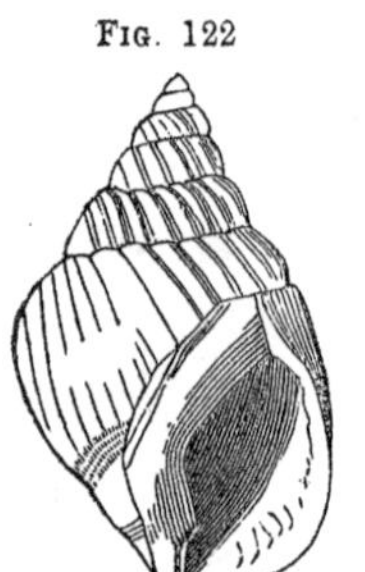

BUCCINUM MULTILINEATUM.—N. S.—(Fig. 124.)

Shell small and thick, turreted; whirls six, and marked by many impressed spiral lines, between which there are also many narrow flat spiral bands; whirls furnished with strong longitudinal ribs, interrupted at the suture, aperture, ovate and less than half the length of the shell; canal short and directed backwards; the body whirl has about thirteen ribs. Rare in the miocene of Cape Fear.

FIG. 123. FIG. 127. FIG. 170. FIG. 125. FIG. 124.

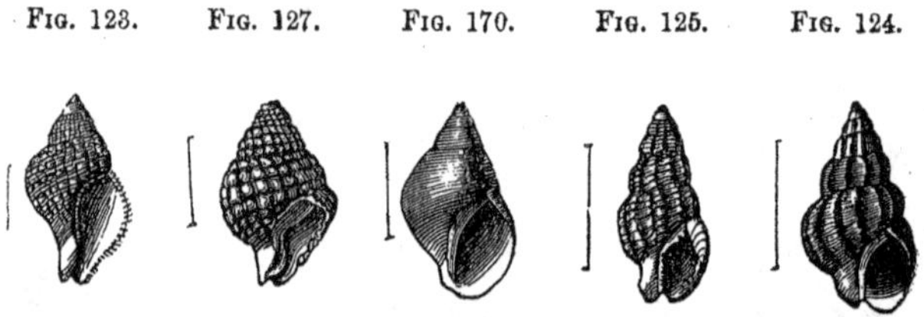

BUCCINUM MONILIFORMIS.—N. S.—(Fig. 125.)

Shell small, thick and robust, rugose; whirls about six, and ornamented with moniliform ribs. This shell, though small, has all the marks of being mature. The flat spiral bands, which as they cross the ribs and give them a beaded appearance, are strongly marked on all the whirls. Rare in the miocene of Cape Fear.

BUCCINUM BIDENTATUM.—(Fig. 126.)

Shell quite small, thick, robust; whirls about five, two upper smooth, the others are ornamented with ribs and spiral bands; aperture oval, acute behind, outer lip furnished with two rather prominent teeth, or short ridges; canal wide and very short.

FIG. 126.

BUCCINUM OBSOLETUM.—(Fig 127.)

Surface granulated; spire shorter than the body. The common species of the coast; is rare in the miocene of North-Carolina. The specimen figured was a young shell, and broken.

FIG. 128.

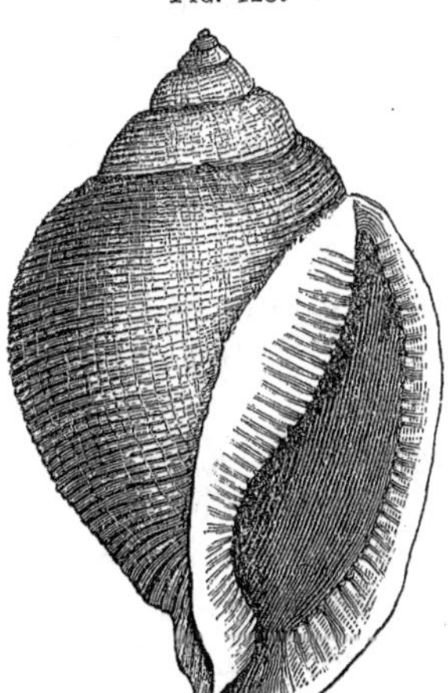

GALEODIA HODGII.—CON. (Fig. 128.)

Shell rather thick; elliptical, obtuse; whirls about five, inflated, and ornamented with numerous fine spiral lines, which are quite prominent at base; these, with the fine lines of growth, give the surface a cancellated appearance; collumellar lip marked with many irregular plicae; aperture nearly twice the length of the spire. Miocene of Cape Fear.

TEREBRA DISLOCATUM; SAY.—ACUS DISLOCATUM.

Shell thick, clongated, acute; whirls many, slightly convex, upper portion constricted, forming a revolving band, parallel to which, there are numerous spiral raised lines; lines of growth longitudinal and conspicuous, which give to the surface a reticulated appearance.

Common in the miocene marls of North Carolina.

18

TEREBRA UNILINEATA; TUOMEY AND HOLMES—FOSSILS OF SOUTH-CAROLINA.—(Fig. 129.)

FIG. 129.

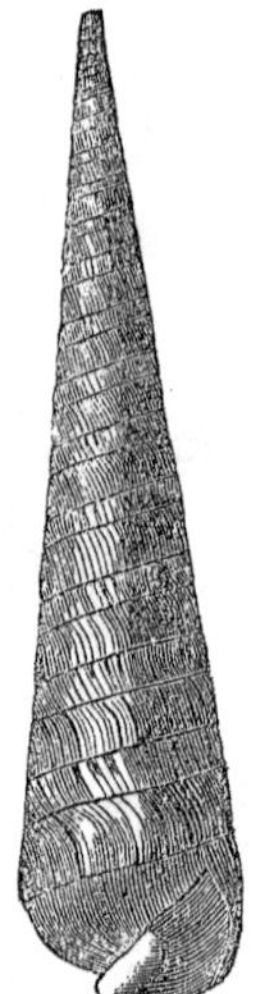

Shell thick, elongate bands alternate, acute, tapering gradually to a point; whirls many, seventeen or eighteen, and ornamented by revolving impressed lines, and passing just above the middle of the whirl; the upper part of the spire is also marked by short longitudinal ribs, which are interrupted by spiral lines. Oblique lines of growth are usually conspicuous. In old specimens, the ribs are obsolete.

Common in the miocene of North-Carolina.

TEREBRA NEGLECTA.—N. S.

Shell terete; spire composed of many whirls, traversed spirally by a deeply impressed line, dividing it into two unequal parts; the lower has three or four interrupted spiral lines, the upper, none. The ribs of the upper part are more obtuse than the lower, and die out before they reach the dividing impressed line; in the lower, they cross it from line to line.

In this species, the revolving lines are fewer than in the T. dislocatum, and in the latter, they are common to both parts of the whirl. In the unilineata, there is but one distinct revolving line.

DOLIUM OCTOCOSTATUM.—N. S. (Fig. 129 a.)

FIG. 129 a. FIG. 181.

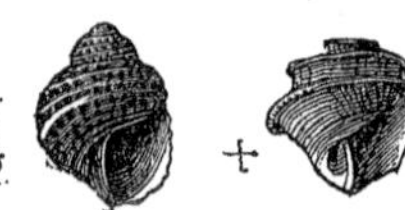

Shell small, thin; whirls three, inflated; body-whirl ornamented with eight spiral ribs, connected by short bars, peristome interrupted; aperture ovate; umbilicus small, open; outer lip crenulated.

OLIVA; LAM.—STREPHONA; BROWN.

The olives are shells of great beauty, being highly polished and covered with a porcellanous enamel, the surface of which is marked by spots and bands of a great variety of colors. The shell is cylindrical, dense and heavy; the spire is short, with channelled sutures, and the aperture long and narrow; the anterior part is notched; the columella is callous and striated obliquely. The body-whirl is furrowed near the base. The olives are the inhabitants of warm climates, and are very active.

OLIVA LITERATA.—SAY. (Fig. 130.)

Shell cylindrical, thick and polished; spire depressed; volutions angular and channelled; apex acute; outer lip sharp, inner marked with numerous sharp folds; aperture linear, incised above and notched below.

This shell is very common in many of the miocene marl beds in the State. It is also living and common on the coast. The fossil frequently retains the polish of the living shell; the colors have disappeared.

OLIVA ANCILLARIAEFORMIS.—LEA.

Shell small, oval, thick, and polished; spire elevated, acute; suture channelled; aperture narrow; inner lip thickened by callus and marked by a few obscure folds.

The foregoing description applies to a small oliva, with a large amount of callus upon its inner lip; but it appears to be a thicker shell than the one to which I have referred it. It is the most common upon the Cape Fear river.

OLIVA.

An oliva, (fig. 131a,) larger than the preceding, and more cylindrical, and having a higher spire, is occasionally found in the miocene beds of the Cape Fear. It has six whirls, and the folds upon the inner lip extend to the suture.

OLIVA CANALICULATA.—LEA.

With this addition to the olives, we have four or five species belonging to the miocene period.

FAMILY CYPREIDAE.

The shells in this family are remarkable for their forms, polish and beauty. They are rolled as a scroll, and are covered with a porcellanous enamel. The spire is concealed, the aperture is long and narrow, and the outer lip is inflexed and thickened. It comprehends the beautiful, spotted and banded shells known as the cowry.

CYPRAEA CAROLINENSIS.—(Fig. 131.)

Fig. 131.

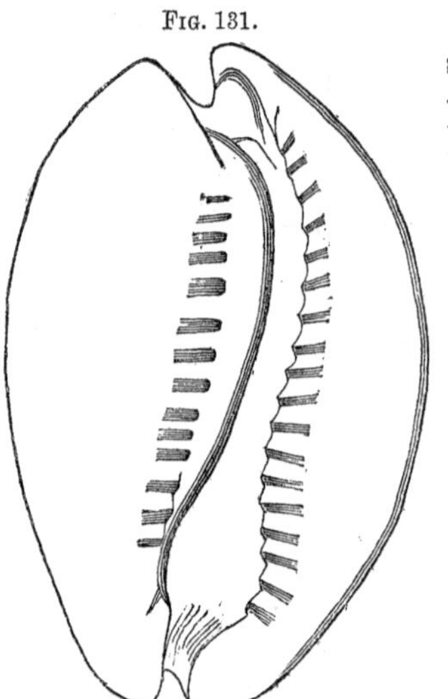

Shell ovate, flattened on the side of the aperture; outer lip prominent at the apex; margins of the lips ornamented with numerous plaits, and receding from each other, beginning at the most prominent part of the whirl. In some of the miocene beds it is quite common.

CYPREA PEDICULUS.

It is a small ovate shell, and transversely ribbed, and with a narrow groove along the back. I have not yet met with it in the marl beds of this State, though it appears to be common in South Carolina.

Fig. 132.

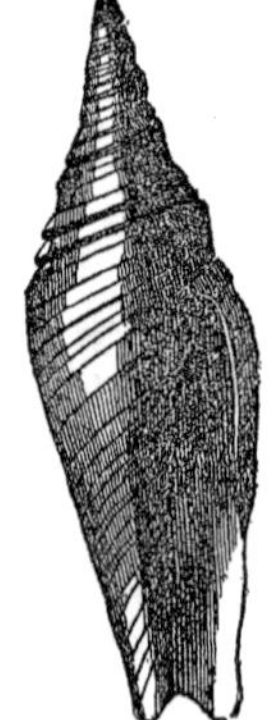

MITRA CAROLINENSIS.—(Fig. 132.)

Shell fusiform, thick, or elongate, and tapering towards each extremity; whirls slightly convex, channeled above, and traversed by numerous spiral raised lines; columella lip, furnished with numerous oblique plaits, of which the upper one is the strongest; canal wide and straight. Miocence marl of North-Carolina. The shell is often found much larger than the figure.

MARGINELLA OLIVAEFORMIS.—PORCELLANA OLIVAEFORMIS: TUOMEY & HOLMES, FOSSILS OF SOUTH-CAROLINA, p. 131.—(Fig. 133.)

"Shell elongated, oval; spire profoundly obtuse; aperture linear; labrum, (or outer lip) tumid, reflexed, profusely crenulated within; columella with three raised plaits."

With this description, several specimens agree, which I have found in the marl beds. It is, however, rare.

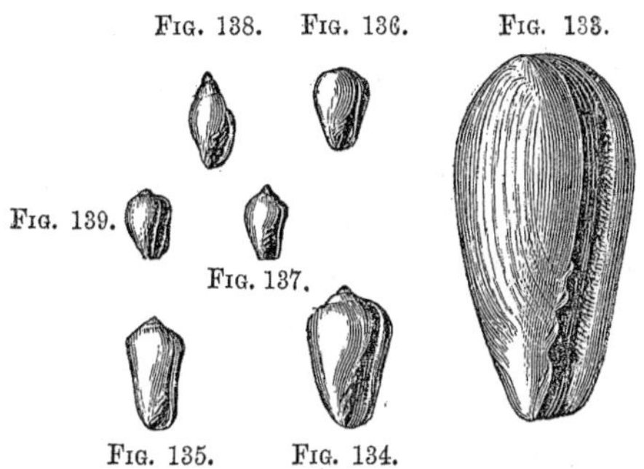
FIG. 138. FIG. 136. FIG. 133. FIG. 139. FIG. 137. FIG. 135. FIG. 134.

MARGINELLA LIMATULA.—(Fig. 134.)

Shell ovate; spire short; outer lip unequally crenulated; columella lip four plaited; aperture contracted above by deposition of callus.

MARGINELLA CONSTRICTA.—N. S. (Fig. 135.)

Shell polished, cylindrical; spire short; aperture constricted in the middle by the imbending of the outer lip; plaits four crowded at the base; margin of the outer lip smooth.

MARGINELLA OVATA.—N. S. (Fig. 136.)

Shell ovate; spire much depressed; aperture uniform; outer lip marked with numerous crenulations within; columella with six or seven plaits, the upper becoming obsolete.

MARGINELLA INFLEXA.—N. S. (Fig. 137.)

Shell oval; spire somewhat elevated; obtuse at base; margin of the outer lip inflexed above the middle; smooth inside; plaits four, and very prominent upon the inner lip. Differs from the constricta in the height of the spire.

MARGINELLA ELEVATA.—N. S. (Fig. 238.)

The thickened outer lip and the plaits of the inner, show this to belong to the genus marginella, though it has a close resemblance to an oliva in the elevation of the spire; whirls four.

ERATO LAEVIS?—(Fig. 139.)

Shell obtusely ovate; wide at the base of the spire; spire depressed; both lips crenulate, but most distinct upon the outer lip; resembles very closely a marginella. Miocene marl of Cape Fear river. (Rare.)

It is difficult to distinguish this from the English species with the aid only of figures. It may be indentical, and I have therefore referred it to the English species.

FAMILY VOLUTIDAE.

The volutes have a thick, short ornamented shell. The spire is particularly so, and it is also provided with a mamillated apex. Aperture is large and elongated, and the columella has several plaits.

VOLUTA MUTABILIS.—CON.

The shell is fusiform and thick, and has a conical spire and a papillated apex; whirls, convex and contracted near the sutures, and the two principal whirls are ornamented with short ribs; lines of growth distinct, and crossed by faint revolving lines; plaits, two and rather distant, and faint indications of an intermediate one. Found in the miocene of the Cape Fear river.

VOLUTA TRENHOLMII: TUOMEY & HOLMES, FOSSILS OF SOUTH-CAROLINA, p. 128.—(Fig. 140.)

"Shell fusiform, ventricose; whirls compressed above, spirally and transversely striated; striae wrinkled and coarse at base; spire short and sub-cancellated, papillated; aperture semi-lunar; outer lip acute, smooth within; columella lip very thin, decumbent, almost obsolete, semi-callous, not distinguishable from the body-whirl, but by outline and color.

Columellar tumid, tortuous; obliquely plaited with three folds."

Fig. 140

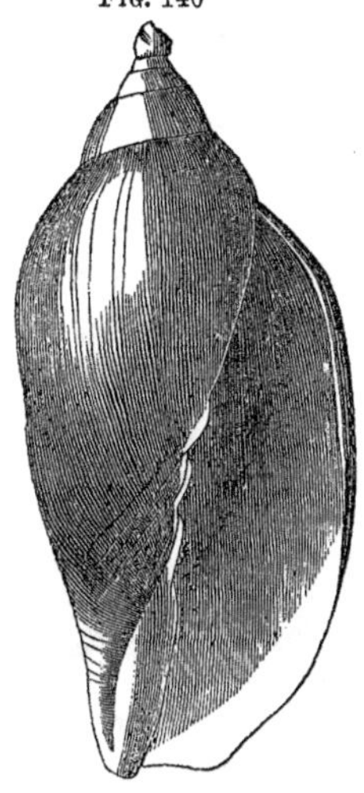

VOLUTA OBTUSA.—N. S. (Fig. 141.)

Shell fusiform, contracted above the body-whirl, and forming thereby a sub-cylindrical spire; spire obtuse apex papillated and hooked; body-whirl plaited longitudinally at its top; columellar lip furnished with only two plaits.

Fig. 141.

Mr. Flowers miocene marl, Bladen county.

FAMILY CONIDAE.

As the name implies, the shells are conical from the great preponderance of the body whirl over the short depressed spire. The aperture is long and narrow, and the outer lip is notched near its suture.

CONUS ADVERSARIUS—CON.—(Fig. 142.)

Shell conical and turned to the left; the surface is marked by revolving lines; towards the face of the pillar lip the lines are strong; whirls of the spire rather concave; edges sub-carinated; labrum sharp, edge convex, and forming a sinus near the suture. Common in all the marl beds upon the Neuse and Cape Fear rivers.

CONUS DILUIVIANUS.—(Fig. 143.)

Shell conical, much smaller than the preceeding, and the whirls are turned to the right; surface markings the same; the revolving lines are less oblique than in the C. adversarius.

They are associated together in about equal numbers. Neither species are found in older beds.

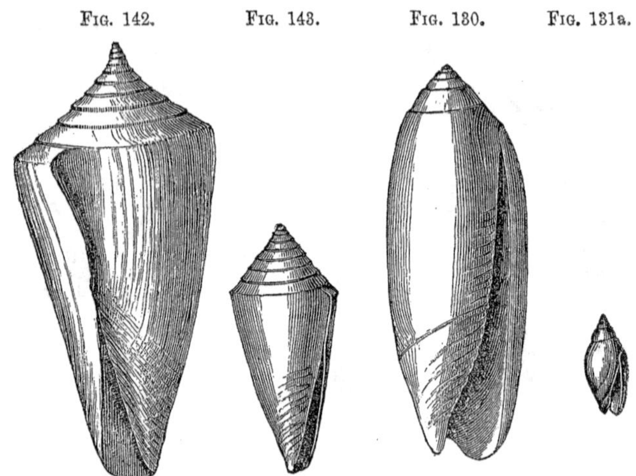

Fig. 142. Fig. 143. Fig. 130. Fig. 131a.

PLEUROTOMA LUNATUM.—LEA. TURRIS LUNATUM.—FOSSILS OF SOUTH-CAROLINA.—(Fig. 144.)

Shell thick, elongate, acute, subfusiform; strongly and obliquely ribbed; spire, eight whirled, angulated above and ornamented by a narrow sutural band.

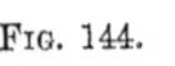

Fig. 144.

The upper part of the whirls are constructed so as to present to the eye a narrow spiral band. Rather common in the marl of Cape Fear river.

PLEUROTOMA LIMATULA.—CON. (Fig. 145.)

Shell rather small, sub-fusiform; spire composed of five or six whirls; whirls constricted above and sub-angulated, forming a sutural spiral collar; ribs oblique and coarse. It is about one inch long.

PLEUROTOMA COMMUNIS.—CON.

Shell small, sub-fusiform; whirls about six, indistinct; body-whirl traversed spirally by four other sharp ridges.

PLEUROTOMA ELEGANS.—N. S. (Fig. 146.)

Shell small, sub-turrited; whirls, about nine, constricted above, ornamented by numerous longitudinal ribs, and traversed by many fine raised spiral lines, which become very distinct upon the pillar lip.

The spiral lines are very regular and equi-distant. The body whirl has about sixteen ribs. Figure natural size.

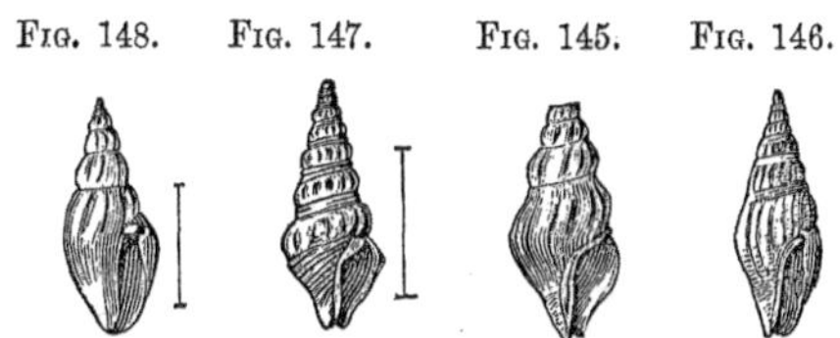
FIG. 148. FIG. 147. FIG. 145. FIG. 146.

PLEUROTOMA TUBERCULATA.—N. S. (Fig. 147.)

Shell small, thick, sub-acute; whirls, seven or eight; apex sub-tuberculated, constricted above, and traversed spirally by rather coarse raised lines; apex papillated, and the first whirl is spirally lined, and without tubercles or short ribs. It is more widely constricted than the preceding.

PLEUROTOMA FLEXUOSA.—N. S. (Fig. 148.)

Shell small, thick, sub-turbinate; whirls, seven or eight, and ornamented by flexuose ribs, which extend across the whirl; ribs alternating with those of the adjacent whirl. There are about ten ribs belonging to the body-whirl.

FAMILY NATICIDAE.

The genus Natica belongs to a family of shells which is characterised by a globular form, few whirls, or a low and obtuse spire, a semilunar aperture, an acute outer lip, and an umbilicus often covered, wholly or in part, by a thick callus. The species are all marine.

NATICA HEROS.—SAY. Fig. 149.

Shell sub-globose, spire depressed, whirls four, convex; lines of growth obscure; aperture, ovate; umbilicus simple and rather large.

This species is common in the miocene marl of North-Carolina. It is also living upon the coast, but is more abundant, according to Dr. Gould, north of Cape Cod than south of it.

NATICA DUPLICATA.—SAY. FIG. 150.

Shell thick, ovate; spire somewhat prominent and pyramidal by the compression of the whirls; and surface marked by faint revolving lines; the lines of growth more distant; umbilicus partially closed by a thick dense callus.

FIGS. 150.

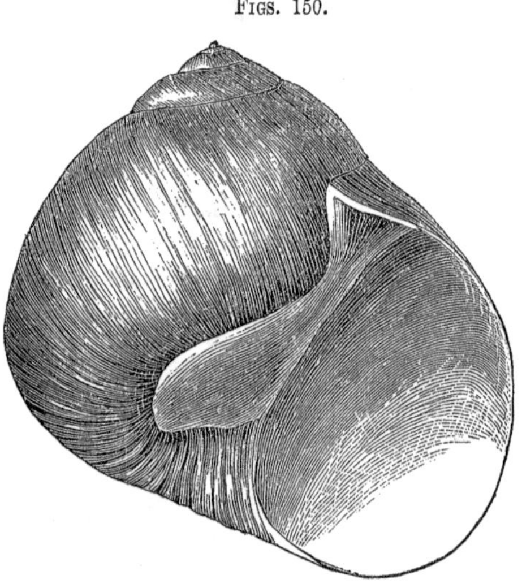

NATICA.—(Fig. 151.)

Shell thick, spire depressed; umbilicus perfectly closed by a thick rough callus, which extends to the angle where it becomes much thickened; suture distinct. It agrees with the clausa in part, but it is a much larger shell, being one inch and eight-tenths in diameter. Fossils answering in size to the clausa exist in the miocene marl on the Cape Fear river.

NATICA CANRENA.—Fig. 152.

Shell rather thick, lines of growth surrounding the spire, very distinct, resembling wrinkles; umbilicus partially closed with callus.

Occurs frequently in the miocene marl of North-Carolina.

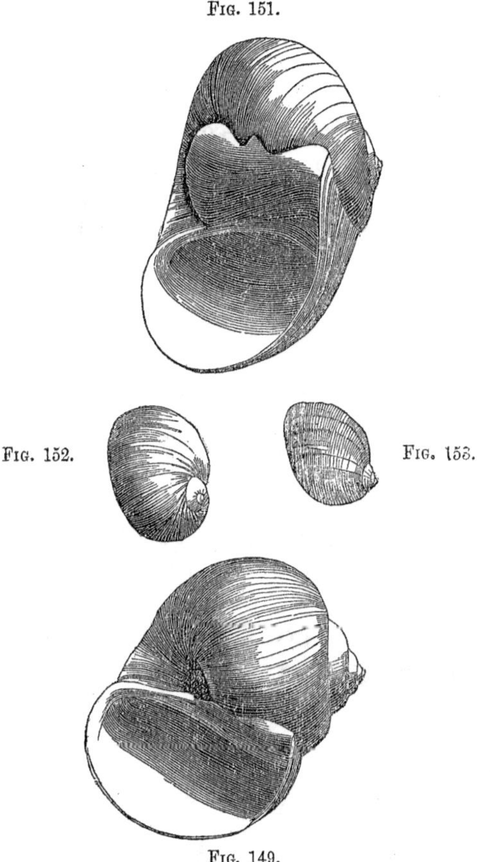

FIG. 151.

FIG. 152.

FIG. 153.

FIG. 149.

NATICA FRAGILIS.—(Fig. 153.

Shell small, surface marked by revolving lines and lines of growth, which give it a cancellated appearance.

FAMILY PYRAMIDELLIDAE.

This family, when restricted to existing species, embraces shells of a small size, and which are spiral slender, pointed and turrited; aperture small, and the columella has one or more prominent plaits. Shells which, in form, bear a very close resemblance to this family, are found in very ancient rocks, but which, in comparison with those of the present day, were of a gigantic size.

PYRAMIDELLA ARENOSA.—CON. (Fig. 154.)

Shell smooth, and still somewhat polished, subulate; suture angularly channelled, columella with two folds; outer lip provided with three teeth. It is a rare shell in the miocene of North-Carolina.

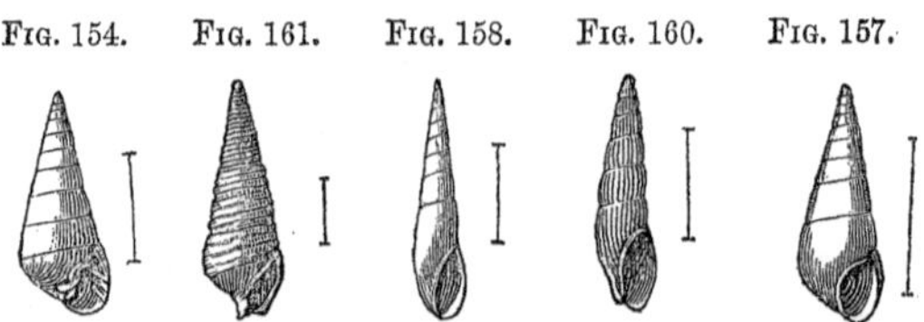
FIG. 154. FIG. 161. FIG. 158. FIG. 160. FIG. 157.

PYRAMIDELLA RETICULTA.—N. S. (Fig. 155.)

Shell turrited; whirls, six or seven, and ornamented by numerous longitudinal ribs, and less distinct spiral lines giving the surface a reticulated appearance; columella three plaited. It closely resembles the P. elaborata—H. E. Lea.

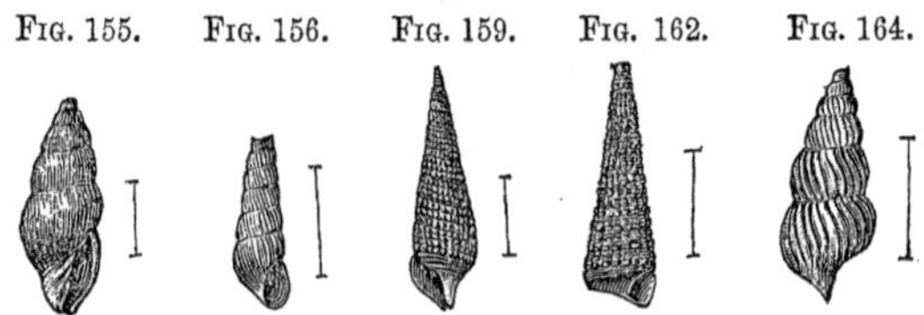
FIG. 155. FIG. 156. FIG. 159. FIG. 162. FIG. 164.

CHEMNITZIA.—(Fig. 156.)

Shell slender, elongated; many whirled; whirls longitudinally plaited and marked by obscure spiral lines; aperture simple, ovate. Rather rare in the shell marl at Magnolia.

CHEMNITZIA RETICULATA.—N. S. (Fig. 156a.)

It has six reticulated whirls, and about six revolving ridges to each whirl. Miocene of Lenoir.

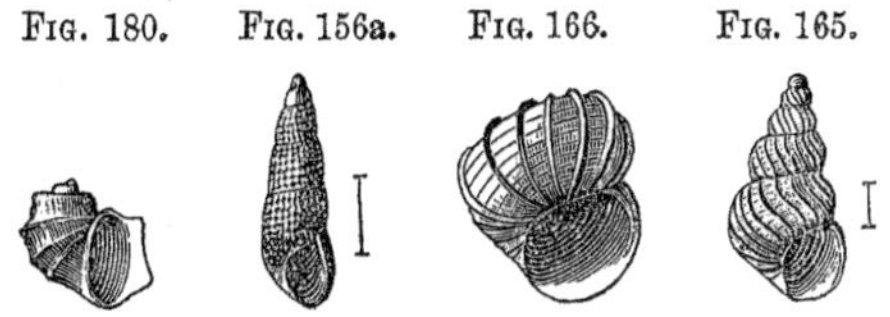

GENUS EULIMA; RISSO.

Shell small, white, polished, porcellanous, elongated, whirls numerous, flat; outer lip sharp, but thickened within; pillar lip reflected over the columella.

EULIMA LAEVIGATA.—PASITHEA LAEVIGATA.—H. E. LEA. (Fig 157.)

Shell small, acute, rather conical, polished and porcellanous; whirls, about nine; suture, obscure linear.

EULIMA SUBULATA.—N. S. (Fig. 158.)

Shell subulate, porcellanous; whirls, nine or ten, slightly convex; sutural space rather wide; aperture elongated. This shell is not uncommon in the shell or miocene marl of Lenoir county.

FAMILY CERITHIADAE—CERITHIUM (TRIPHORIS) MONILIFERUM: H. E. LEA.—(Fig. 159.)

Shell subulate, sinistral, thick, costate, sutures small; whirls, ten, flat; ribs three, moniliform; columella smooth; canal short and deep.

CERITHIUM.—(Fig. 160.)

Shell small, elongated; whirls, many, slightly convex, ornamented with numerous longitudinal ribs, which extend across the whirl; canal short and deep.

CERITHIUM ANNULATUM.—N. S. (Fig. 161.)

Shell small but thick; whirls many, ornamented with three

sharp spiral ridges. These ridges are but slightly oblique to the axis of the shell.

CERITHIUM BICOSTATA.—N. S. (Fig. 162.)

Shell small, thick, tapering from the base; whirls many, and ornamented with two spiral, nodulose ribs.

TEREBELLUM ETIWANENSIS.—TOUMEY AND HOLMES—FOSSILS OF SOUTH-CAROLINA.

Shell subulate; whirls many, pointed, flattened and ornamented with two sharp spiral ribs; sutural line deep, especially below.

This shell presents considerable variation in passing from its immature to its mature state. In the young the spiral ridges are placed near the suture, and the space between is concave; the waving lines of growth gives it an obscurely beaded appearance. It is the most common univalvė in the marl beds of Edgecombe county.

TEREBELLUM CONSTRICTUM.—N. S.

Shell rather thin terete; whirls many convex; lower ones deeply constricted on the line of suture, and ornamented by two principal raised revolving lines placed nearer the lower margin than the upper; the finer parallel lines are numerous; longitudinally, the spire is frequently marked by obsolete ribs; lines of growth indistinct. It differs from the T. Etiwanensis in the position of the principal revolving lines, and the lower rounded whirls.

FIG. 163.

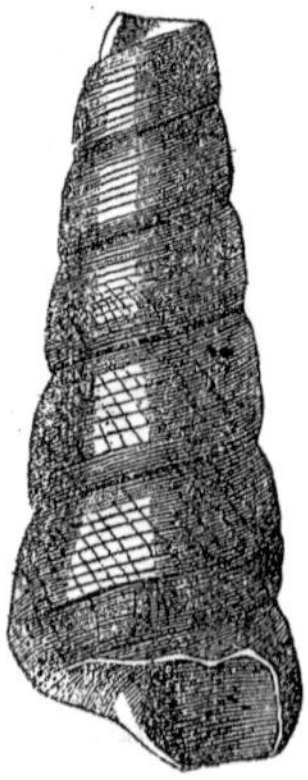

TEREBELLUM BURDENII.—TOUMEY & HOLMES. FOSSILS OF SOUTH-CAROLINA, P. 122. (Fig. 163.)

" Shell subulate, turrited; whirls flattened, spirally ribbed and transversely striated, which give the ribs a beaded character."

SCALARIA MULTISTRIATA.—(Fig. 164.)

Shell, small whirls numerous, rather convex and ornamented with many sharp longitudinal ribs.

All the specimens of this species of scalaria which fell under my observation were imperfect at the aperture. Shell marl of Lenoir county.

SCALARIA CURTA.—N. S. (Fig. 165.)

Shell thin and delicate; whirls about four, ornamented with rather flexuose, sharp, longitudinal ribs. Shell marl of Lenoir county.

SCALARIA CLATHRUS.—(Fig. 166.)

All the specimens of this species, when found, were imperfect. It differs from the preceding in having transverse ribs between the longitudinal ones.

PETALOCONCHUS.—LEA.—PETALOCONCHUS SCULPTURATUS. (Fig. 169.)

Shell vermiform, tubular, provided with two longitudinal plates internally; externally it has nodulose ribs or costae. The shell is curiously twisted into knots, but sometimes it is rolled up into a coil somewhat conical, as in the figure, after which it is coiled irregularly. It is very common in the miocene marl beds of the State.

FIG. 169.

FAMILY LITORINIDAE.—LITORINA LINEATA.—N. S. (Fig. 170.)

Shell rather small, thick conical; whirls five nearly flat, and the two lower are ornamented with many spiral ridges, which are crossed by obscure lines of growth; three upper whirls smooth.

FAMILY TURBINIDAE—TROCHUS PHILANTROPUS.—(Fig. 167.)

FIG. 167.

Shell conical, but rather depressed; whirls slightly angular at base, and ornamented with spiral beaded lines, alternating in size.

TROCHUS.—(Fig. 168.)

FIG. 168.

It appears to differ from T. armillatus, but I am unable to refer it to any of the species described in the miocene beds.

DELPHINULA QUADRICOSTATA.—N. S. (Fig. 180.)

Shell small, thin; whirls, few, angulated and furnished with four ribs, which are crossed by lines of growth; aperture angular.

Found occupying the interior of the large univalve shells of the miocene.

ADEORBIS.—WOOD. (Fig. 181.)

I have placed this figure under this genus, though it does not agree with it in every particular.

FAMILY TORNATELLIDAE.

This family has a convoluted shell; it is cylindrical, or subcylindrical, with a long narrow aperture; columella plaited.

TORNATINA CYLINDRICA.—N. S. (Fig. 182.)

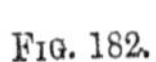

FIG. 182.

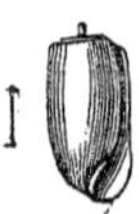

Shell small, convoluted, cylindrical, porcellanous, or polished; spire depressed; whirls, angulated; suture channelled; aperture long and narrow; outer lip arcuate; columella with one fold.

This small shell resembles a cyprea, or some of the smaller species of olivas. It is not uncommon in the miocene; it is usually found in the cavities of the larger univalves.

FAMILY HELICIDAE—LAND-SHELLS.—HELIX TRIDENTATA.
(Fig. 183.)

Shell depressed, or flattened, convex; whirls, four and obliquely wrinkled; aperture contracted and furnished with two teeth on the outer lip, and one upon the inner lip; the latter is curved.

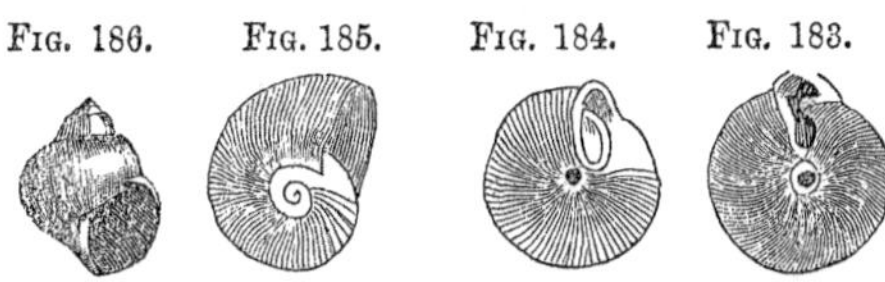
FIG. 186. FIG. 185. FIG. 184. FIG. 183.

H. LABYRINTHICA.—(Fig. 184.)

Shell small and of a conical form; whirls, six and marked with oblique lines of growth; lip reflexed; inner lip furnished with a single tooth extending within the shell.

FAMILY LIMNEIDAE.—FRESH-WATER SHELLS.—PLANORBIS BICARINATUS.—(Fig. 185.)

Shell deeply concave on both sides; whirls, three; carinated on both sides; lip on the left extending beyond the plane of the preceding whirl.

This fresh water shell is rare in the miocene beds of the Cape Fear.

FAMILY PALUDINIDAE.

This family embraces certain gasteropods, most of which live in fresh water, as lakes, ponds and rivers. The form of their shells is conical or globose, covered with a thick green epidermis. The aperture is rounded and the whirls convex; peristome continuous.

PALUDINA SUBGLOBOSA.—N. S. (Fig. 186.)

Shell rather thin, turbinated; whirls, four, rounded or convex, short; aperture rounded; third whirl marked by four or five spiral obsolete lines. It has a close resemblance to Gould's and Halderman's genus Amnicola.

Miocene of Cape Fear, but it is by no means a common shell.

CLASS BRACHIOPODA.

ORBICULA LUGUBRIS.—CON. (Fig. 187.)

FIG. 187.

Shell corneous, oblong-ovate, depressed; concentrically lamellose; apex behind the centre; posteriorly, it is marked by a few radiating lines; interiorly, it is smooth, and there is a short longitudinal ridge on the median line. In some of the miocene beds in Wayne county, it is quite common.

FAMILY DENTALIDAE.—TOOTH SHELLS.

The dentalidae are hollow, curved tooth-like shells. They are usually ornamented by longitudinal ridges, but sometimes they are smooth and polished. They have a round or circular aperture.

DENTALUM ATTENUATUM.—SAY. (Fig. 188.)

Shell gently curved, and ornamented with twelve rounded ribs; aperture circular. Common in the shell marl of this State.

FIG. 188.

FIG. 190.

FIG. 189.

D. THALLUS.—CON. (Fig. 189.)

Shell small, polished, curved and tapering towards both extremities. Common in the shell marl.

CAECUM ANNULATUM.—N. S. (Fig. 190.)

Shell minute curved; ends subequal; aperture circular; surface annulated.

This minute shell is quite common in the miocene of this State. It is found in the interior of larger ones, which it probably inhabited.

FAMILY CALYPTRAEIDAE.—LIMPETS.—BONNET LIMPETS—CUP AND SAUCER LIMPETS.

The limpets have but one valve. It is sometimes saucer

shaped or sub-conical, and passing into a cone, within which there is an appendage somewhat similar in form to the outer cone. These cones are frequently sub-spiral. They adhere to rocks and stones with their apertures below.

CRUCIBULUM COSTATUM.

Shell rather thick, circular at base, and furnished with strong but rather irregular ribs; apex sub-central; margin crenulated.

CRUCIBULUM RAMOSUN.—CON. (Fig. 191.)

Shell ovate; apex sub-central; ribs prominent and ornamented by a series of subordinate diverging ridges, but partially interrupted by the lines of growth; inner cup sub-conical, entire, and marked by circular ridges, or lines of growth.

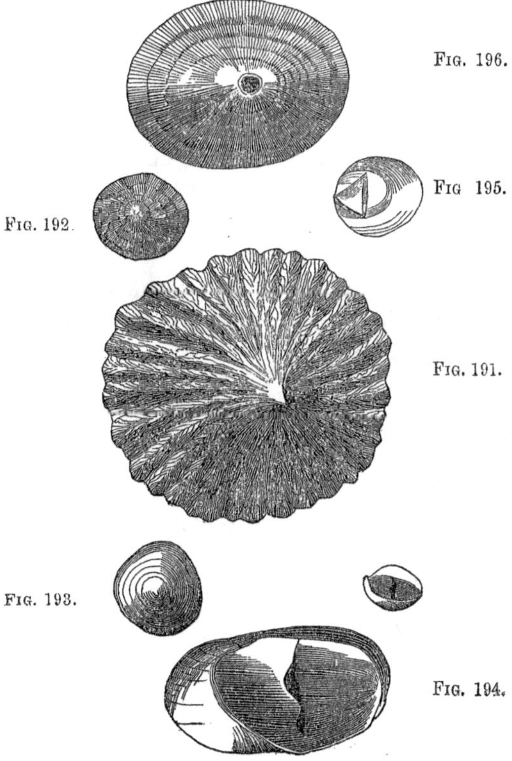

FIG. 196.

FIG 195.

FIG. 192.

FIG. 191.

FIG. 193.

FIG. 194.

C. DUMOSUM.

Shell depressed, sub-conical, oblong or oval at base; surface ornamented with spiral ribs, and whose spines are hollow.

C. MULTILINEATUM.—(Fig. 192.)

Shell rather small, depressed, very thin; apex elevated, sub-central, disk marked with radiating lines. Rather common in the miocene. Usually occupies the interior of other shells.

TROCHITA CENTRALIS.—(Fig. 193.)

Shell rather small, very thin, round, ovate; apex medial minutely spiral and acute. Associated with the foregoing shells of this family.

CREPIDULA.—LAM.

Crepidula has the limpet shape, but a posterior oblique marginal apex. Interior has a horizontal plate, forming a partition which curves the posterior half. They vary in form, which is very much dependent upon the surface to which they are attached.

CREPIDULA FORNICATA.—(Fig. 194.)

Shell obliquely oval; surface convex, smooth or wrinkled; apex turned to one side: diaphragm concave below, occupying half the shell. Common in the miocene of North-Carolina.

CREPIDULA SPINOSA.

Shell depressed, oval, costate and spinous, especially towards the margin. Common in the miocene.

CREPIDULA PLANA.—SAY. (Fig. 195.)

Shell nearly flat, slightly convex; diaphragm convex; the form is very variable, assuming the shape of the surface upon which it rests.

FAMILY FISSURELLIDAE.—KEY-HOLE LIMPETS.

Shell limpet shaped; some have the margin notched in front; in others the apex is perforated. Adhere to rocks and stones.

FISSURELLA REDIMICULA.—(Fig. 196.)

Shell ovate, oblong, elevated, and rather thick; surface ornamented with fine longitudinal ridges, which are intersected by circular lines of growth, which gives the surface a reticulated appearance; margin entire, but ridged internally; apex truncated, figure inclined, oblong.

This shell is not an uncommon occupant of the shell marl beds of this State.

CLASS LAMELIBRANCHIATA.

FAMILY OSTREIDAE.

"Shell inequivalve and nearly inequilateral; free or adherent resting on one valve; beaks central, straight ligament internal; muscular impression single and behind the centre; hinge usually without teeth."

OSTREA VIRGINIANA.

Shell thick, strongly and radiately plicated; concentrically laminated and imbricate; upper valve nearly flat; pliated towards the margin; beaks laterally curved; very variable. Common in the miocene beds of North-Carolina.

OSTREA CAROLINENSIS.

Shell ob-ovate, thick, compressed, concentric lamina imbricated, and transversely plaited; beaks broad and prominent. Fosset large and bounded laterally by strong ridges.

Occurs in the miocene of North-Carolina, but is less common than the preceding.

Ostrea radians and O. sellaeformis belong also to the miocene beds, together with the Anomia ephippium; the latter is always broken.

EXOGYRA COSTATA.—(Fig. A.)

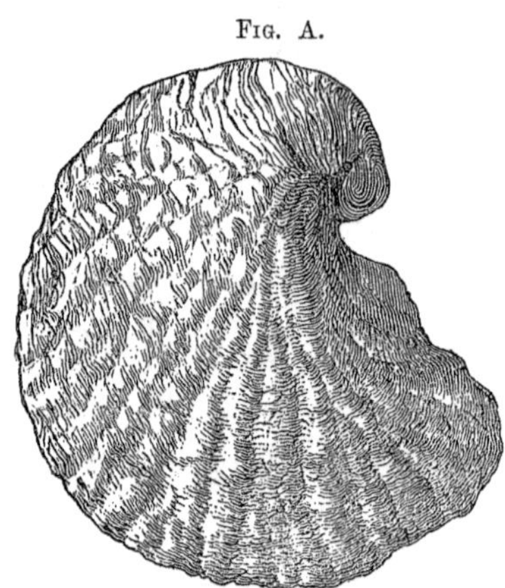

Fig. A.

Shell sub-oval, very thick, lower valve convex, and covered with strong corrugated ribs; apex lateral, with about two volutions; upper valve flat, thick, supplied with numerous elevated concentric squamose plates. It belongs to, and is, one of the characteristic fossils of the green sand at Black Rock, on the Cape Fear, and at Rocky Point, twenty miles north of Wilmington. It is found in the miocene at several places on the Cape Fear, but its presence is due to accident.

CUCULLAEA VULGARIS.—(Fig. B.)

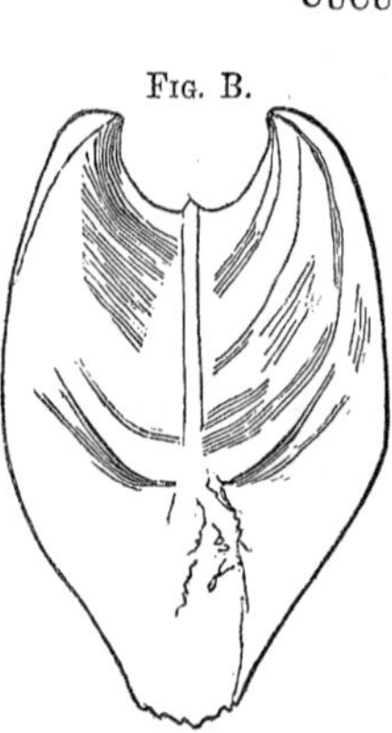

Fig. B.

This fossil occurs in the form of an inside cast of the shell; it is inflated, subtriangular, flattened before, beaks prominent and in-curved; shell thick, and marked with numerous delicate longitudinal striae.

It is associated with the Exogyra and Belemnitella at Black Rock in the green sand.

The C. vulgaris is placed here from its association with the E. costata.

FAMILY PECTENIDAE.—PECTEN, SCALLOP.

Shell sub-orbicular, regular, resting on the right valve, usually ornamented by fretted or scaly ribs radiating from the hinge; right valve most convex, with a notch below the front ear; hinge margin straight, united by a narrow ligament; cartilege internal in a central pit.

The scallop of our coast is regarded as a delicacy. It lives in shallow water, and is taken in great numbers at low tide from banks which are just submerged. They move through the water by opening and shutting their valves. Fossil pectens or scallops are very abundant in most of the miocene marl beds in this State. The large scallops, P. Jeffersonius and P. Madisonius abound in beds upon the Neuse and Tar rivers, while they are less numerous upon the Cape Fear. Another large species is found upon the Meherrin, in Northampton county, which I have not met with elsewhere. It replaces the English species, the Pecten princeps, which it closely resembles.

PECTEN COMPARILIS.

Shell medium size; both valves convex with twenty-three or twenty-four ribs, prominent and angular inside at base; ribs and spaces between nearly equal; ears radiately striate. One of the most common fossils upon the Cape Fear.

PECTEN EBOREUS.—(Fig. 197.)

FIG. 197.

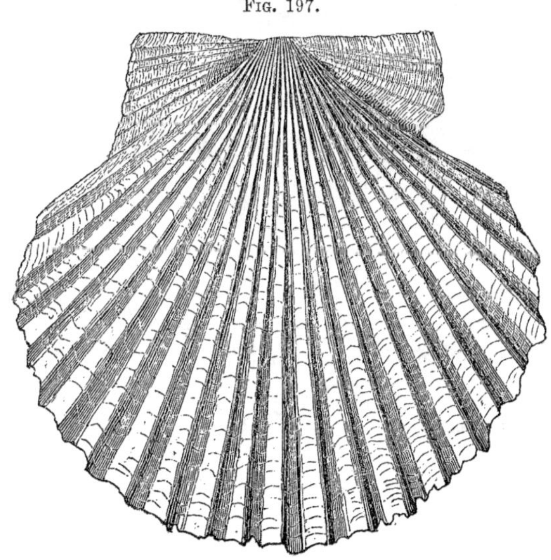

Shell comparatively thin, and light and compressed valves; circular, sometimes oblique and equilateral; ribs twenty-four, marked on the outside with concentric squamose lines of growth, which are undulating, the last of which are strong; lower valve less convex than the upper. It differs from the comparilis in being concentrically marked, and thinner, besides it grows much larger.

PECTEN PRINCEPOIDES.—N. S.—(Fig. 198.)

Shell large, rather thick, compresed, sub-inequilateral, radiating striae coarse and very numerous; transversely marked

FIG. 198.

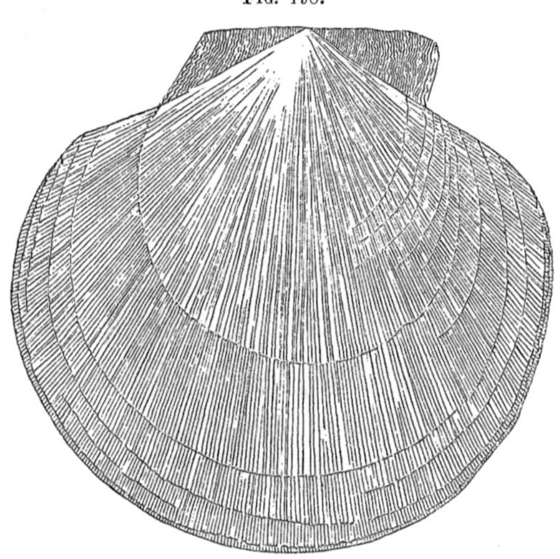

by lines of growth, giving the surface a wrinkled appearance; ears unequal; buccal ear sinuate, radiating striae numerous, inside smooth, striae obsolete; fig. reduced.

This is a large species of pecten, is closely allied to the P. princeps of the English crag. It is common in the miocene marl on the Meherrin river, at Murfreesboro'. It is five inches long, and five and a quarter wide. It is readily distinguished by the absence of ribs proper, and the presence of coarse radiating striae, which have intermediate ones, which do not reach the hinge or umbo; many of the striae, however, fork or divide.

P. PEEDEENSIS.

Shell thick and strong, broadly ovate; ribs, eight, broad striae, lines of growth strong towards the margin; beak projecting beyond the hinge line.

Only one valve has been found of this species, and being old and its striae obliterated in part, and its characters are less distinct than is usual in specimens belonging to this genus.

P. MORTONI.

Shell large, circular, compressed, thin, pearly; equivalve equilateral; concentrically marked by fine lines of growth; on the outside, ribs are invisible; inside, ornamented by about eighteen pairs of ribs, which are prominent at the margin, and obsolete towards the hinge.

This beautiful shell occurs in the miocene at Waccamaw Lake, North-Carolina, and has not been observed upon the Neuse or farther north.

P. JEFFERSONIUS.—(Fig. 199.)

FIG. 199.

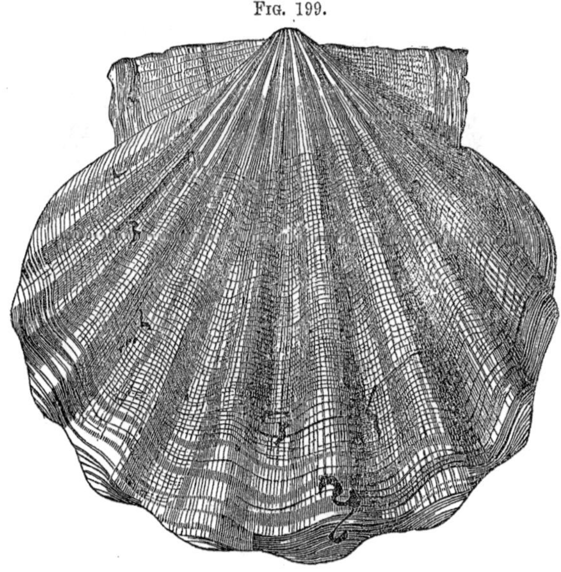

Shell very large, ribs, ten, and wide, and longitudinally marked by fine ridges, which are not squamose. This species is sometimes between nine and ten inches wide, and seven or eight inches long, and are often used in cooking oysters in place of a frying pan. It is one of the characteristic fossils of this miocene.

P. MADISONIUS.—(Fig. 200.)

In the P. Madisonius, the ribs number about fifteen, and they are ornamented with three squamose ridges each. There is also an equal number between them; they coalesce towards the hinge.

Fig. 200.

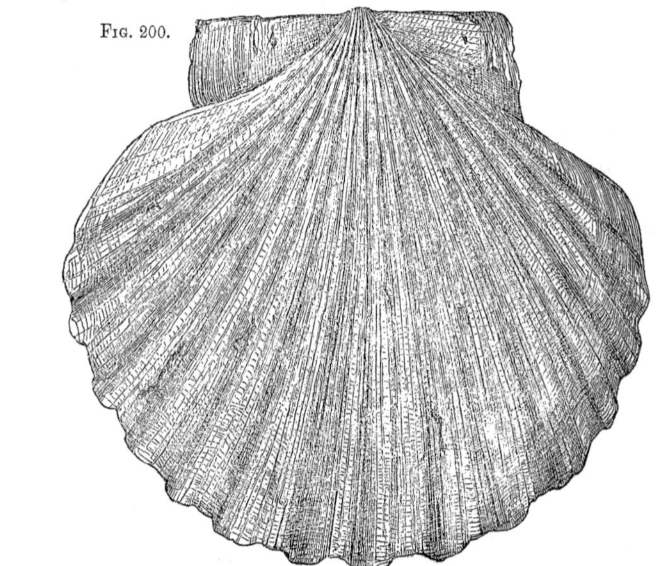

Fig. 201.

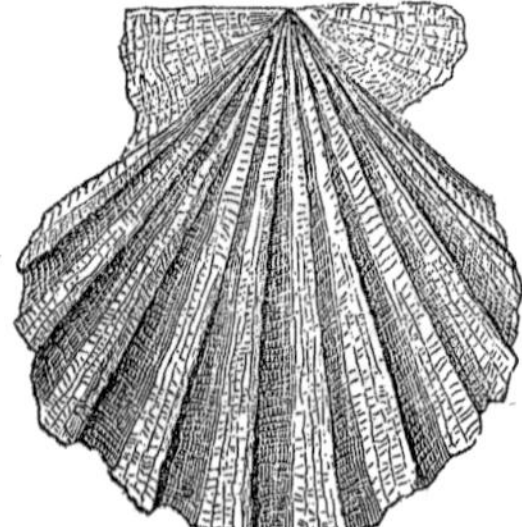

A. pecten, (fig. 201,) is quite common in North-Carolina, which I have not been able to refer to its proper species. It is one of the most common in the shell marl of the middle part of the eastern counties. It has ten prominent ribs, but they are ornamented in a different style from that which prevails in the young of the P. Jeffersonius.

One of the most common pectens of the white eocene marl, is represented by figure 202. It differs from the P. membranacea in having only about half the number of ribs. The P. membranacea having upwards of eighty, while this has about forty-four.

FIG. 202.

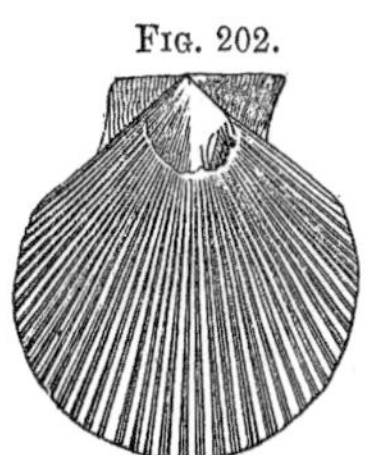

An observer cannot fail to perceive the striking difference in the species of pectens of the white eocene marl of New-Hanover and Onslow counties, and those of the miocene.

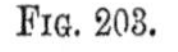

FIG. 203.

PLICATULA MARGINATA.—(Fig. 103.) Shell strong and thick, but rather small; valves sub-equal, ovate, wedge-form, with three strong radiating plicae.

FAMILY MYTILIDAE.—MYTILUS INCRASSATUS. (Fig. 203A.)

Shell nacreous, thick, somewhat inflated, marked with concentric lines of growth; anterior margin arched accuminate; posterior rounded, somewhat dilated; umbones acute. It is usually much injured by exfoliation and rarely perfect.

FIG. 203A.

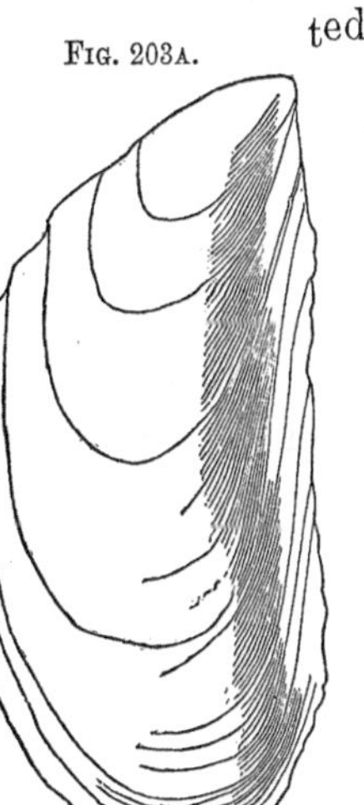

CRENELLA.—(Fig. 203B.)

Shell small, short, thin, smooth in the middle; hinge, margin crenulated behind the ligament. It appears to be rare, though it may be owing to its frailness. Miocene.

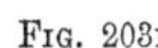

FIG. 203B.

ARCADAE.

The valves in the Arcadae are equal, regular, and usually oblique; the teeth are arranged in long rows, resembling a comb; at the extremes they are longer and frequently curved or corrugated.

ARCA LIENOSA.—SAY.—(Fig. 204.)

Shell large, inflated, oblique; ribs subequal, numerous, with a groove or channel in the middle; anterior side angu-

FIG. 204.

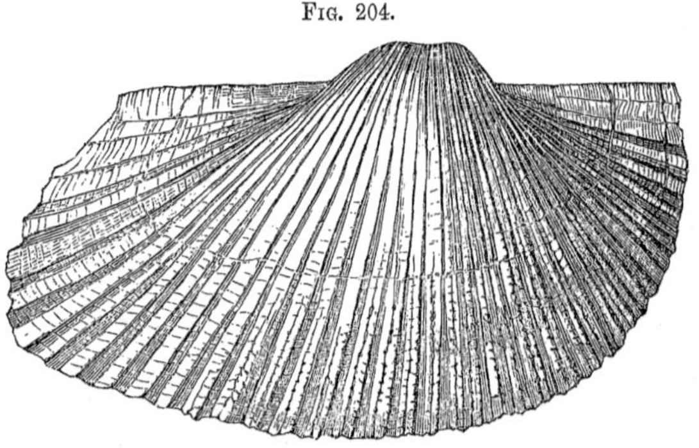

lar; lines of growth distinct, giving a striate appearance; the ligament area is marked by strong lines diverging from beneath the umbo; umbones distant; inside margin strongly sulcate or ribbed. It has about 37 ribs. A living shell upon the Florida coast, but found abundantly in the miocene of North-Carolina.

A. SCALARIS.

Shell oblong, ovate; ribs twenty-one, strong and transversely rugose, ligament area short, transversely marked by lines, and crossing striae parallel to the hinge line.

A. INCILE.—SAY.

Shell very oblique, sub-quadrangular; anterior side very short, posterior sinuate; ribs unequal, stronger on the posterior margin; rounded before, angular behind, and much produced; umbones incurved, distant; ligament area crossed by transverse lines.

This shell has about thirty-one principal ribs, with intervening raised lines, and transversely marked by lines of growth.

A. CENTENARIA.—(Fig. 205.)

Shell sub-quadrate and ovate, nearly straight and slightly

FIG. 205.

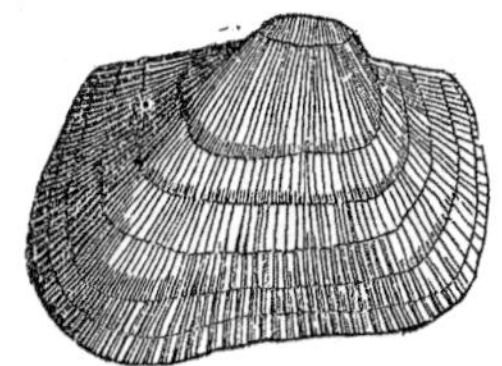

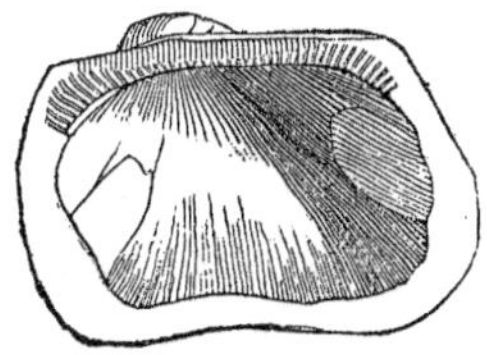

contracted at base; ribs fine, alternating in size; margins rounded; beaks approximate; hinge area narrow; margins entire.

The striae or ribs in this species are very numerous and fine, while these together with its quadrangular form will serve to distinguish it from others of the same genus. Common in the miocene of North-Carolina. The figure was drawn from a specimen obtained from the indurated sand beneath the miocene bed at Elizabethtown, Bladen county, and is referred to the centenaria but with doubt.

A. IDONEA.

Subcordate inequivalve ventricose; elongated and only slightly oblique; beaks very prominent and distant; ribs about twenty-five, crenulated, or transversely ridged; hinge area wide and marked by divergent striae or channels. Common in the miocene of North-Carolina.

A. TRANSVERSA.

Shell rather thin, subrhomboidal, rounded with about thirty-two ribs; area rather narrow, with two or three undulated grooves. Common in the miocene, and still living upon the coast. A. limatula and stillicidium are also miocene shells, and common in the marl beds of the Cape Fear river.

VERTICORDIA.—WOOD.—(Fig. 206.)

Fig. 206.

I have met with two or three specimens only of the fossil which I have referred to this genus. It is found in the interior of large shells.

GENUS PECTUNCULUS.

Shell orbicular, nearly eqilateral, smooth and radiately striated; hinge with a semi-circular row of transverse teeth.

PECTUNCULUS SUBOVATUS.—(Fig. 207.)

Shell orbicular, inequilateral, with radiating sulci, becoming obsolete with age; teeth nearly obliterated in the centre; teeth largest on the shorter side of the valve; marginal ones broad and separated;—Conrad. This is probably one of the most common miocene fossils of the shell marl in the State.

Fig. 207.

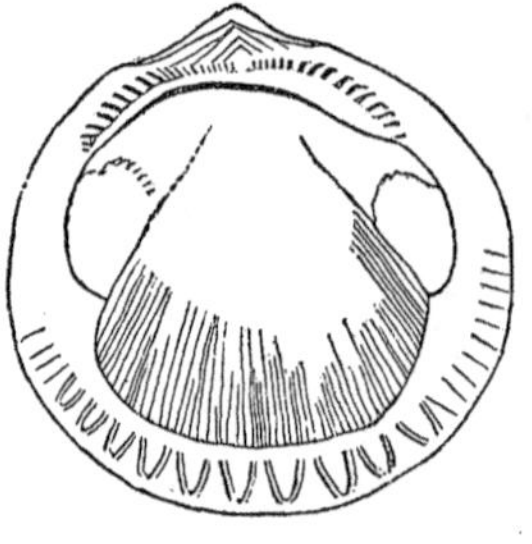

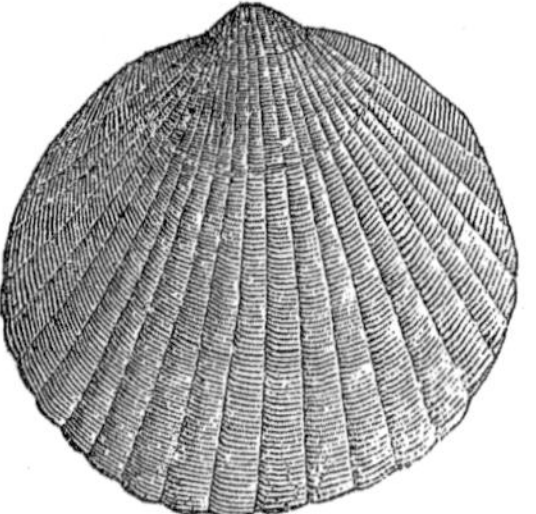

P. LENTIFORMIS.

Shell orbicular, sub-equilateral; the radiating striae are numerous; beaks small in proportion to the size of the shell; hinge teeth in the centre, wanting or obsolete. This fine species in some marl beds upon the Cape Fear, is quite common, and is very large and thick; some are four to four and a half inches across.

P. ARATUS.—(Fig. 208.)

FIG. 208.

This is the smallest species of this genus belonging to the shell marl. It is also one of the most common. P. passus and P. quinquerugatus are also common in certain localities.

LEDA ACUTA.—(Fig. 208A.)

FIG. 208A.

Shell small, thick, inflated posteriorly; margin acute or beaked, slightly open; anterior margin, short rounded; surface concentrically striated. This fossil resembles nucula, but it is not pearly in the interior, and its abdominal margin is smooth.

NUCULA PROXIMA.—(Fig. 208B.)

FIG. 208B.

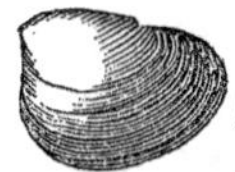

Shell small, ovate, smooth, interior pearly; anterior margin short; posterior side elongated, obtuse; margin crenate. N. limatula is more common in the marl beds of this State than the N. proxima; miocene.

FAMILY CHAMACIDAE.

The shell is thick, inequavalve, with sub-spiral beaks, hinge teeth 1—2, muscular impression one, and large; reticulated palleal line simple.

CHAMA.

The shell is attached to other bodies by its left umbo; hinge-tooth of the free valve thick, curved, and received between the teeth of the other valve.

CHAMA ARCINELLA.—(Fig. 209.)

Shell thick, or orbicular-cordate squamose; the radiating ribs spinose, strong, tubular or folded; intervening space coarsely punctate and rugose. Common in the marl bed at Elizabethtown, Bladen county.

Fig. 209.

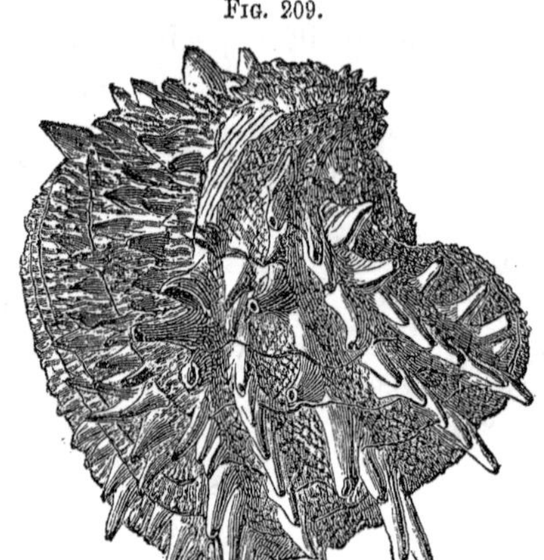

CHAMA CORTICOSA.—(Fig. 210.)

Fig. 210.

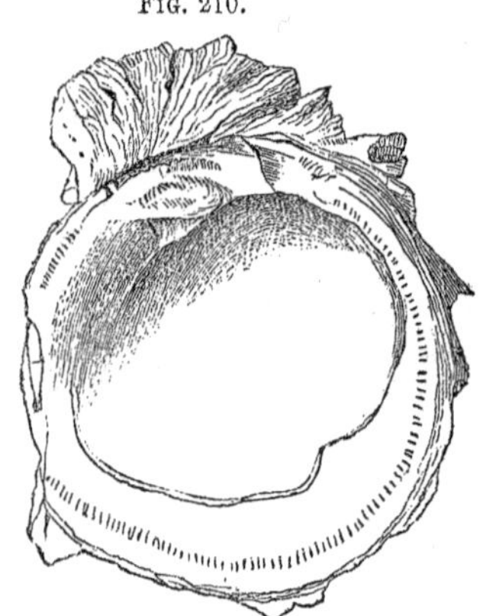

Shell thick, squamose, or concentrically laminated and imbricate; lamina striated, sinistral, crenulated interiorly; upper valve flat. Figure lower valve natural size. Abundant in the miocene of North-Carolina, especially on the Cape Fear.

CHAMA CONGREGATA.

Shell thick, orbicular, with its surface composed of plates or lamina; in the flat valve the plates are crenulated or plaited.

CHAMA STRIATA.—N. S. (Fig. 211.)

Fig. 211.

Shell small, ovate, rather thick for its size, lower valve distinctly striate. Usually found in the hollow or inside of the univalves.

FAMILY CYPRINIDAE.

Shell regular, equivalve oval or elongated; valve close, solid; epidermis thick and dark; ligament external, conspicuous cardinal teeth 1—3 in each valve; pedal scars close to or confluent with the adductors; pallial line simple.—Woodward.

ASTARTE

Shell small, thick, compressed, smooth or concentrically furrowed; lunule impressed; ligament external; hinge teeth 2—2; anterior tooth in the right valve large and thick.

ASTARTE CONCENTRICA.—(Fig. 212.)

Fig. 212.

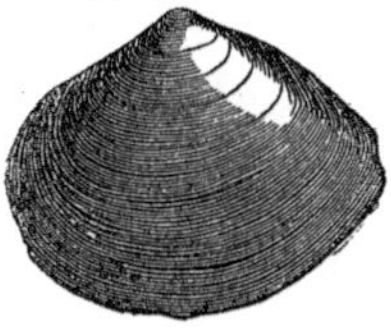

Shell small, thick, triangular, compressed, concentric; furrows close and regular; umbones acute, recurved; margin crenate. It is about one inch long, and one broad. It is rather common in the miocene of North-Carolina.

ASTARTE UNDULATA.—(Fig. 213.)

The broad, variable and concentric furrows will serve to distinguish it from the foregoing. It is comparatively a broader shell. The Undulata seems, however, to be quite variable, and the figure shows one of the extremes of this species.

Fig. 213.

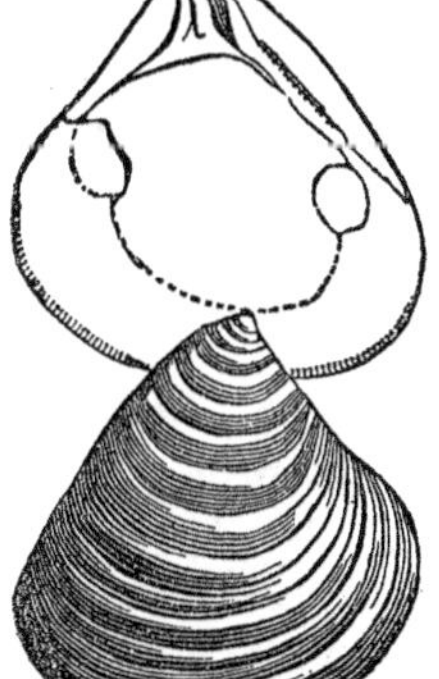

CRASSATELLA UNDULATA.—(Fig. 214.)

Shell oblong, ovate, compressed, marked upon the outside with coarse concen-

Fig. 214.

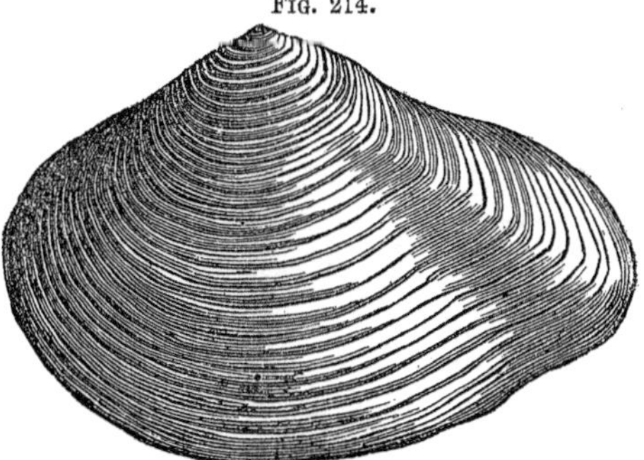

tric striae; umbo flattened; apex sub-acute; inner margin entire. One of the most common fossils of the shell marl.

C. GIBBESII: TUOMEY & HOLMES, FOSSILS OF SOUTH-CAROLINA; p. 74. (Fig. 215.)

FIG. 215.

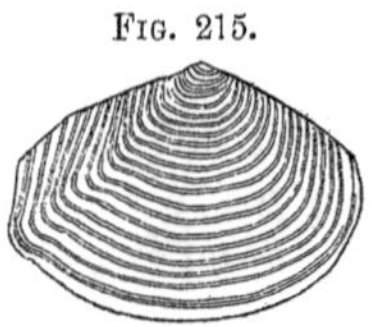

"Shell somewhat triangular, thick, concentrically furrowed; buccal side rounded; anal side somewhat beaked, angular, with a longitudinal ridge; umbones incurved; lunule somewhat excavated."

In addition to the foregoing, I may add the following as common in the North-Carolina shell marl beds: Crassatella alta, C. Marylandica, C. Protexta, C. Melina.

FAMILY CYCLASIDAE.—CORBICULA DENSATA.—CYRENA DENSATA. CON.—(Fig. 215A.)

FIG. 215A.

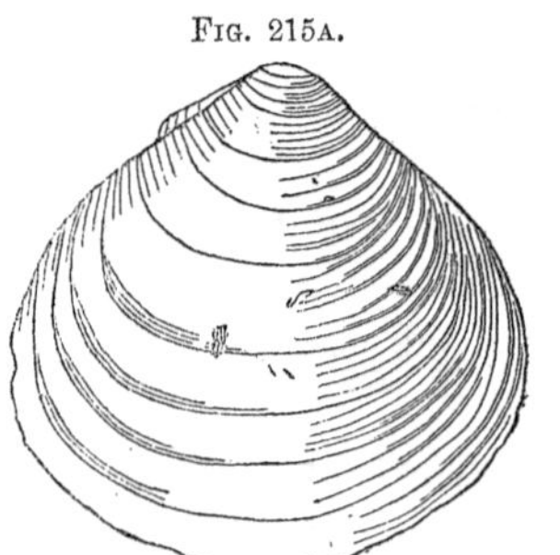

Shell orbicular striated concentrically, polished, lateral teeth elongated.

This shell is very abundant at the miocene marl bed of Mr. Flower, on the Cape Fear.

FAMILY CORBULIDAE.—CORBULA CUNEATA.—(Fig. 215B.)

FIG. 215B.

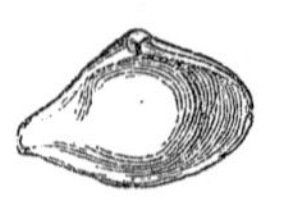

Shell small, thick, ovate, concentrically striate; anterior margin rounded; posterior elongated, or somewhat rostrate. Common in the shell marl.

FAMILY LUCENIDAE.

This family have orbicular shells, both free and closed with hinge teeth, somewhat variable as one or two laterals, or one and one, and the other obsolete; pallial line simple, muscular impressions two, elongated and rugose. The family is principally composed of tropical and temperate species, and live upon sandy or muddy bottoms, and exist from the sea shore or shallow water to the greatest habitable depths.

LUCINA BRUGIERE.

The shell is orbicular, white, with depressed umbones, and the margins are either smooth or only finely crenulated; hinge teeth 2—2, laterals 1—1, muscular impressions rugose; anterior, elongated and within the pallial line; umbanal area with an oblique furrow.

LUCINA PENNSYLVANIA.—LINN. (Fig. 216.)

Shell orbicular, thick, solid, and concentrically ribbed, or posteriorly it has a strong fold or groove. The fold extends across the shell, and produces a notch in the pallial margin. Common in the miocene upon Neuse and Cape Fear rivers.

Fig. 216.

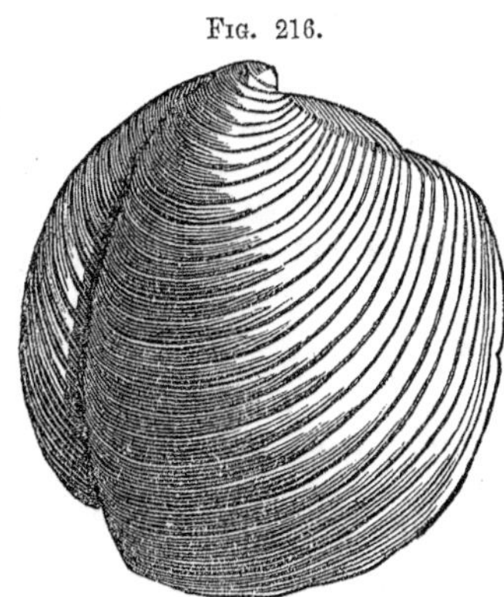

LUCINA CONTRACTA.

Shell orbicular, somewhat inflated; ribs concentric, unequal, marked in the intevals with striae; posteriorly the margin is channeled.

It is larger than the preceding, and has no fold, and its ribs are unequal.

L. CRENULATA.—(Fig. 217.)

Fig. 217.

Shell small, thin, orbicular, somewhat inflated, concentrically lamellated, lunule excavated. Common.

In addition to the foregoing, the following species have been observed in the miocene: Lucina anadonta, L. radians, L. divaricata, L. multihineata, and L. squamosa.

FAMILY VENERIDA.

This important family is represented by many existing species in our seas at the present time. It is too well known to require a minute description. It is, however, known from other forms by its regular oblong thick shell, though it is sometimes nearly round; by its strong external ligament, and its three diverging prominent teeth in each valve. Its pallial line is sinuated.

The venerida are elegant and beautiful shells, often highly colored, though some of the best known are externally dull. This family appeared first in the Oolitic period, and they have increased in number and importance down to the present time, when they have acquired their maximum developement.

VENUS MERCENARIA.

Shell solid, surface marked by numerous concentric lines of growth, obliquely cordate; posterior margin produced; anterior short; umbones recurved, lunule cordate; pallial line sinuated; margin crenulated.

VENUS TRIDAENOIDES.—CON. VENUS DIFFORMIS.—SAY.

Shell very thick and heavy; globose, wrinkles upon the surface undulating; plaits wide, extending from the umbo to the margin.

This species may be distinguished by its thickness and wide external plaits, which are usually strongly marked, though sometimes they are feebly developed. It is one of the most common fossils of the miocene beds of North-Carolina.

VENUS RILEYI.

Shell large, thick, oblong, posterior margin prolonged, anterior one short; surface concentrically striate, and marked by fine, longitudinal lines, which are distinct after the dermal covering exfoliates. This is one of the largest species, being sometims 6—7 inches wide. Common in the miocene of Cape Fear river.

V. CRIBRARI—CON. (Fig. 218.)

FIG. 218.

Shell thick, medium size, slightly ventricose, furnished upon the outside by about twenty-five sharp lamelliform concentric and recurved ribs, crenulated upon the umbonal side; ribbed or ridged transversely on the ventral side, the ridges extending across to the adjacent rib; lunule crenulated.

Recent upon the coast of North-Carolina.

V. LATILIRATA CON.—VENUS PAPHIA.—LAM. (Fig. 219.)

FIG. 219.

Shell sub-trigonal, thick and ponderous for its size; ribs fine, concentric, and very thick; irregularly stirate, crenulate upon the lower margin; umbo slightly flattened.

This shell is readily known by its thick ribs, and deep subci between them. Common in the miocene of North-Carolina.

VENUS MELTASTRIATA.—(Fig. 220.)

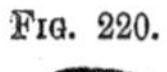
FIG. 220.

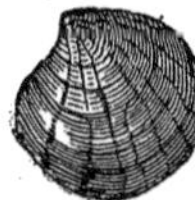

Shell small, sub-orbicular, striated concentrically, rather irregular, interruptedly radiated.

Venus pramagna, cancellata and subnasuta are also rather common fossils of the miocene.

CYTHERCA SAYANA.—(Fig. 221.)

Shell inflated, concentrically striate, anterior side angulated; umbones prominent, incurved; margin smooth; lumule cordate.

FIG. 221.

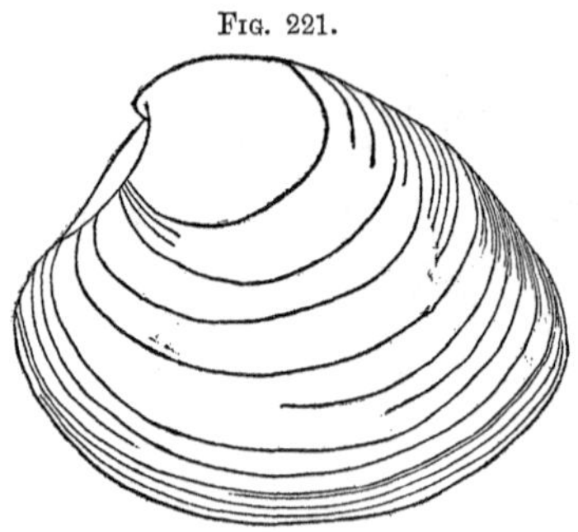

C. REPOSTA.—(Fig. 222.)

Shell smooth, moderately inflated, thick, beaks prominent, dorsal margin depressed; anterior margin rounded, lunule lanceolate.

FIG. 222.

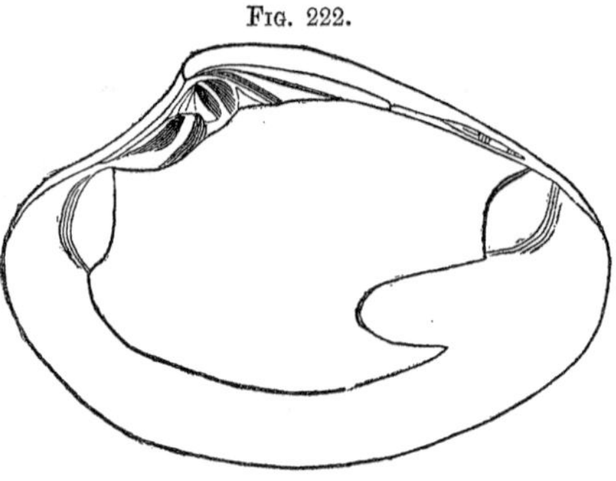

C. REPORTA.—(Fig. 223.)

This fossil, which the annexed figures represent, is very common in a sandy marl bed in Brunswick county. It preserves its original polish, and closely resembles the foregoing. It is, however, proportionally wider than the repostia. It is highly polished and smooth, but has concentric striae. Umbones flattened, the flattened part extending across the shell, being bounded anteriorily with an obscure rounded ridge.

Fig. 223.

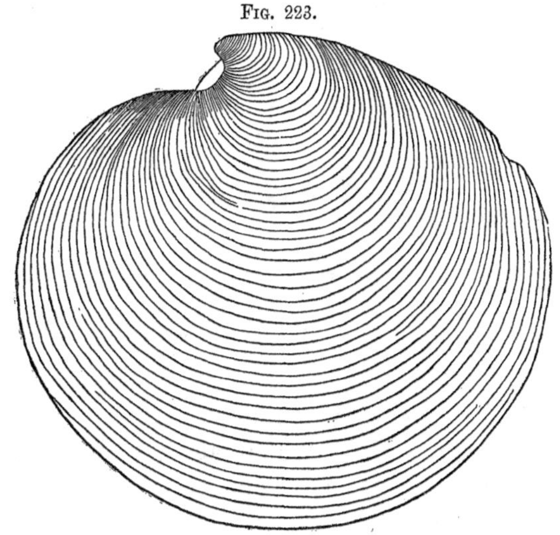

ARTEMES TRANSVERSUS.—N. S. (Figs. 223a and 224.)

Fig. 223a.

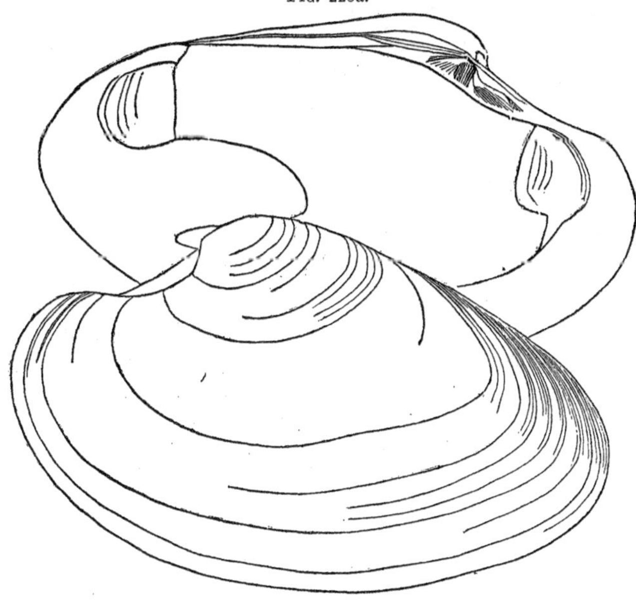

Shell sub-orbicular, depressed, sub-equilateral, concentrically striate; broader than long; lumule small, lines of growth or concentric striae regular, simple, and somewhat coarse and distant. Fig. 224 shows the hinge.

Fig. 224.

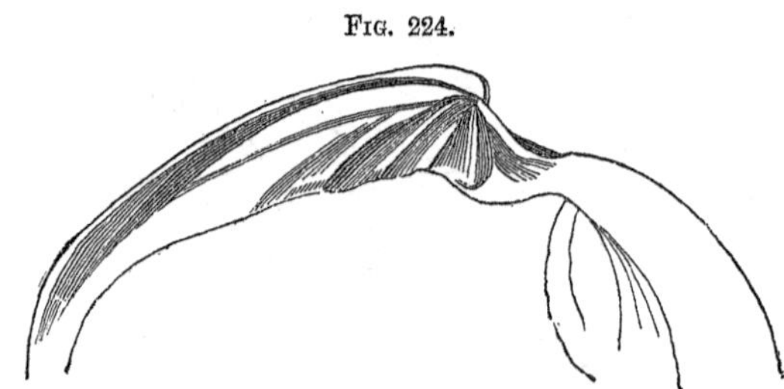

This fossil appears to differ from the Artimus concentrica of the coast; its linus of growth are about half as numerous and are also continuous from one margin to the other, excepting a few on the anterior margin.

In the living coast species the lines of growth are less regular, and coalescent near both margins; it is orbicular also, being as long as wide. The fossil, however, closely resembles the living one of the coast, though it differs as much from it as Artemis acetalubum of Conrard.

Species which belong to the miocene and which remain undescribed: A. acetabulum, A. concentrica.

FAMILY TELLIMIDAE.—TELLINA BIPLICATA.—CON. (Fig. 225.)

Shell rather large, thin, sub-oval, inequivalve, sub-ventricose, marked with rather obscure radiating lines, and impressed with an oblique fold in each valve. The remaining species of Tellina belonging to the miocene are T. Alternata, T. Polita, and T. Flexuosa.

FIG. 225.

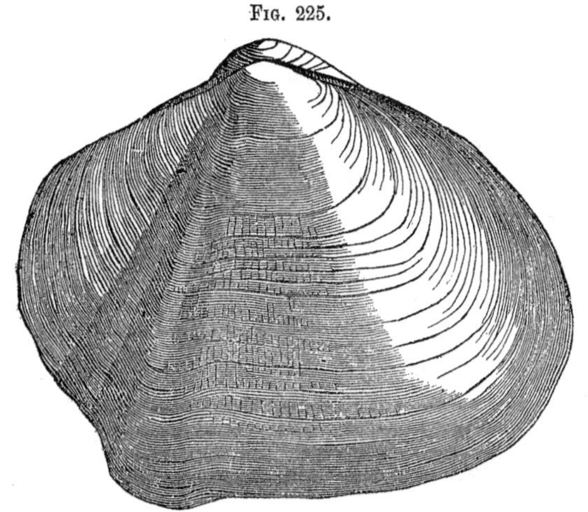

TELLINA LUSORIA.—(Fig. 225A.)

Shell oblong, narrowed posteriorly, slightly gaping or reflected; pallial sinus deep; concentrically striate; posterior margin marked with one or two folds; surface still brown; concentric striae are in the form of raised sharp lines, not impressed lines of growth. The Tiphonal inflection is in contact with the pallial line, in which respect it agrees with P. Sammobia, but its hinge teeth are 2—2 in both valves.

FIG. 225A.

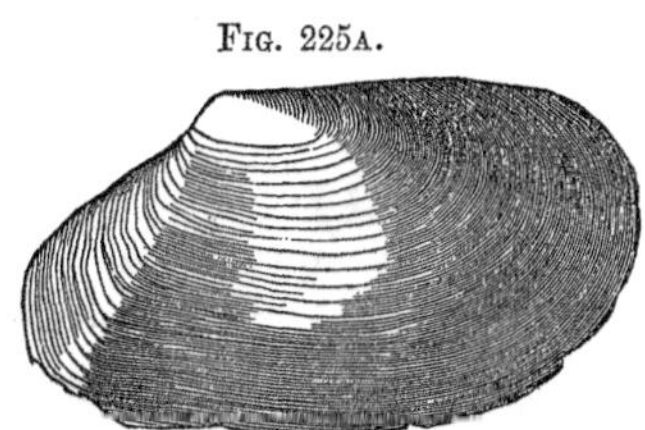

GENUS DONAX.

"The general form is trigonal, or wedge form, valves closed, front produced, posterior short; margins usually crenulated; hinge teeth 2—2; laterals 1—1 in each valve; pallial sinus deep."

DONAX.—(Fig. 226.)

Fig. 226.

Shell triangular, rather abruptly truncate behind, and traversed by a ridge from the umbo to the base; surface marked by obscure radiating lines; base crenulated. This small shell differs from the *variabilis* in its proportion; it is more triangular, and is not produced so much in front.

Donax Variabilis probably occurs in the marl of North-Carolina, but has hitherto been overlooked.

FAMILY MACTRIDAE.—GENUS MACTRA.

"The shell is equivalve, and nearly equilateral; the anterior hinge tooth is in the form of an inverted A; lateral teeth doubled in the right valve."

MACTRA CONGESTA.

Shell rather small, but thick at the umbo; triangular, rather inflated; inequilateral; rounded anteriorly, and posteriorly it is produced. Very common in the marl of Wayne and Edgecombe.

MACTRA LATERALIS.—SAY. (Fig. 227.)

Fig. 227.

Shell small, rather thin, smooth, sub-triangular; lines of growth fine; posterior side elongated, or margins sub-equal, rounded before; umbo rather prominent. A very common fossil of the miocene.

MACTRA SIMILIS.—SAY.

Shell thin, of a medium size, margins sub-equal, concentric, striae very fine, at intervals deep, beaks nearly central. The living ones of the coast have a longitudinal rounded ridge running from the beaks to the base and obscure radiating lines, though only visible in a favorable position.

GNATHODON GRAYII.—(Fig. 226a.)

Shell rather thick, sub-triangular, inflated, inequilateral, anterior margin rounded; posterior elongated or wedge form. Rather common in the shell marl beds of Cape Fear.

Fig. 227a.

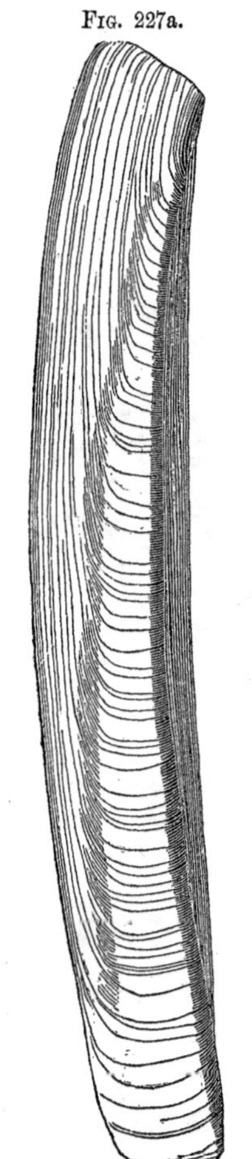

Fig. 226a.

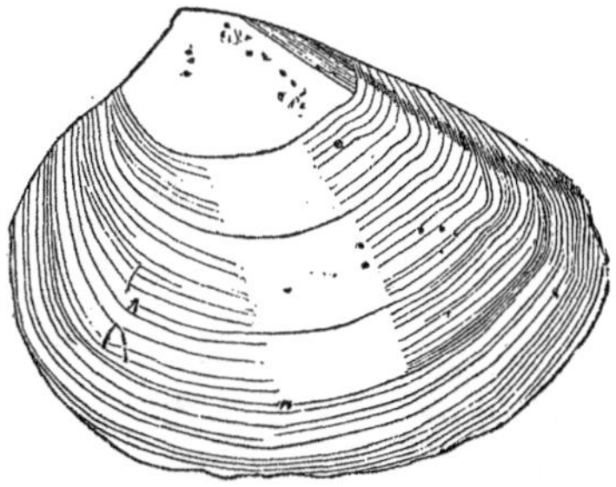

FAMILY SOLENIDAE.—SOLEN ENSIS.

(Fig. 227a.)

This common shell of the coast is sword shaped, with the anterior and posterior margins truncate.

SOLECURTIS SUBTERES.—CON. (Fig. 228.)

Shell rather small, thin, somewhat sword shaped; anterior and posterior margins rounded, ventral margin concave, or arched.

Fig. 228.

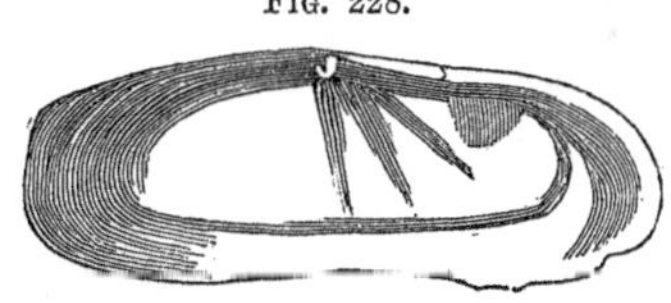

P. CARIBOEUS.—(Fig. 228a.)

Is common in the miocene, but the valves are rarely entire. I should, however, express some doubt respecting the identity of the specimen figured with this species.

Fig. 228a.

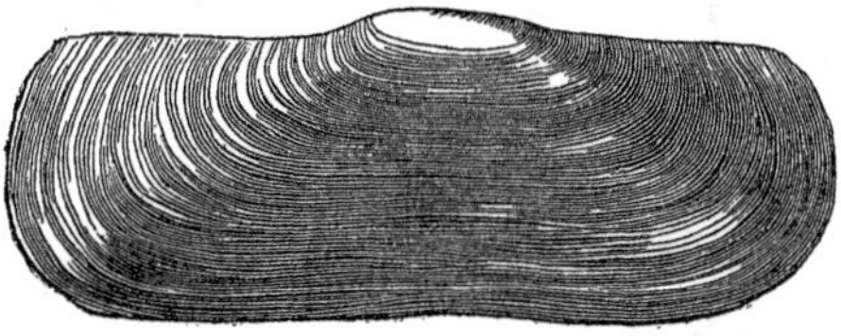

FAMILY ANATINIDAE.—PANOPEA REFLEXA. (Fig. 229.)

Shell large, thin, oblong, ovate; wrinkled and margin gaping widely and reflected. Common in the shell marl of Edgecombe county.

FIG. 229.

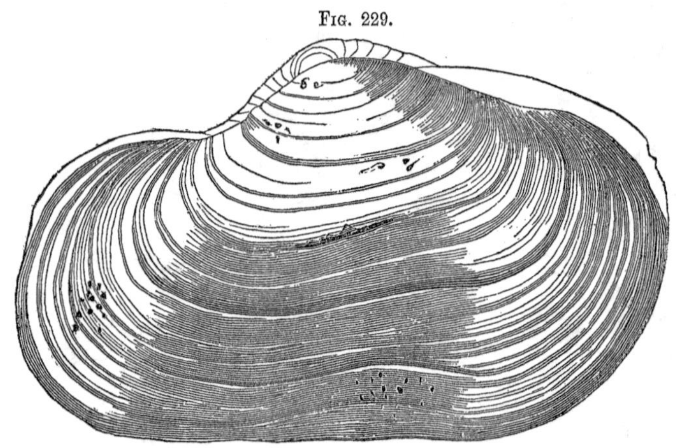

PHOLADOMYA ABRUPTA.—(Fig. 231.)

Shell oblong, oval, substance nacreous; surface ornamented with from three to five radiating ridges. This beautiful bivalve is quite common in a marl bed in Edgecombe county but rarely entire.

FIG. 231.

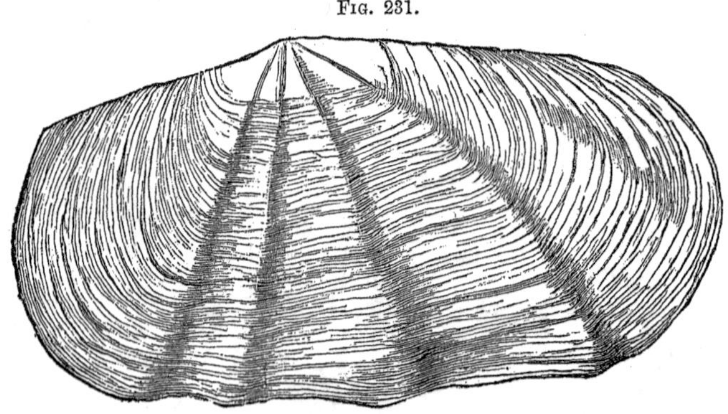

FAMILY PHOLADIDAE.

These species of Pholas have been found in the miocene of this, viz: P. Costata, P. Oblongata, and P. Memmingeri. They are rarely if ever entire, but their fragments are not uncommon.

FAMILY CARDIDAE.*—CARDIUM MAGNUM.—CARDIUM VENTRICOSUM.

Shell large, inflated, obliquely cordate, radiately ribbed, ribs flattened, anterior ones crenulated.

This magnificent fossil is found occasionally in the miocene. It is quite common in the pliocene, and is now very abundant upon the coast, near Beaufort.

CARDIUM MURICATUM.—(Fig. 232–'3.)

The specimen given in the figure resembles the muricatum,

FIG. 232–'3.

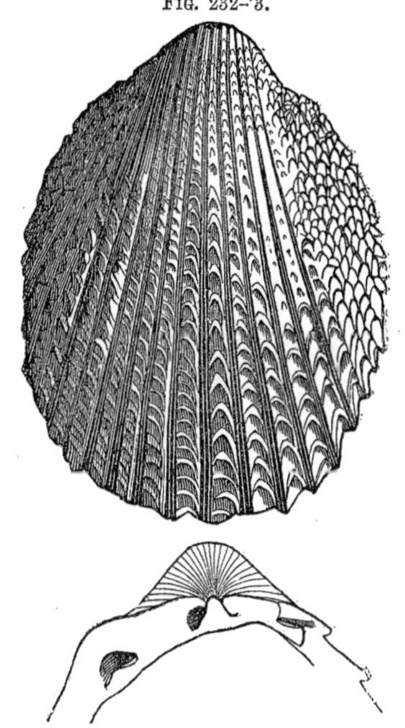

* The families *cardidae* and *carditidae* should have preceded veneridae.

but it is more elongated, and its crenulations appear to differ. I have obtained only one specimen; and hence, cannot speak of the permanence of its characters. It occurs in Walker's Bluff, on the Cape Fear.

Cardium sublineatum is a common fossil of the Cape Fear and Neuse marl beds.

FAMILY CARDITIDAE.—CARDITA ARATA.—(Fig. 234.)

FIG. 234.

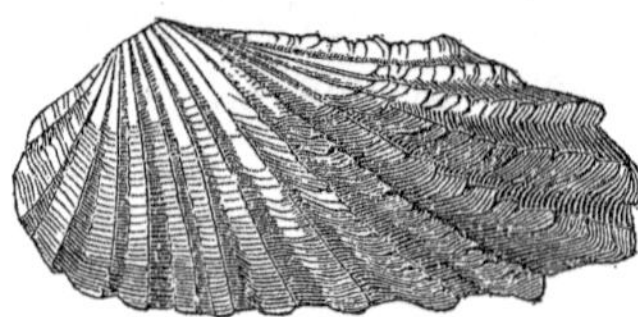

Shell rather thick, oblong, and ornamented with fifteen or sixteen elevated scaly ribs; anterior side very short; posterior margin oblique; inner margin crenate.

C. PERPLANA.—(Fig. 235.)

Shell small, rather thick, triangular, inequilateral, radiately ribbed, striated; posterior side produced, anterior short. Common.

FIG. 235.

FIG. 236.

FIG. 236. A.

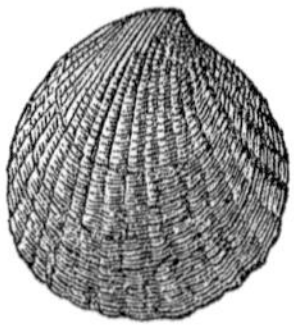

C. ABBREVIATA.—(Fig. 236.)

Shell small, thick, triangular, oblique; ribs strong and crenate; umbones acute. Common.

CARDITA TRIDENTATA.—(Fig. 236. A.)

Shell round, triangular, thick; ribs strong and crenulate; beaks turned forward; valves with two teeth in the left, and one in the right valve.

CARDITA CARINATA.

Shell small, thick, wide on the abdominal side; ribs strong and radiating; muricated; anterior side short.

CHAPTER XIX.

RADIATA.

Considerations relative to animals belonging to this type.—Aberant forms of the Echinodermata.—Species described.—Bryozoa, Polyparia, etc.

Echinodermata comprehends a class in the Kingdom, Radiata, whose organization belongs to the stellate type. This sub-class derives its name from the character of the integument, and its appendages, which remotely resemble that of the *hedge-hog*. Some are called sea-urchins, others star-fishes. In most of the families of this great class, the integument is protected by calcareous spines. The integument itself is coriaceous, but it takes into its composition a large quantity of lime which imparts to it firmness and durability. The skin is complicated in its structure. It is made up of an immense number of plates of a polygonal form. They amount to 600 pieces in all. These are dove-tailed together in the most perfect manner, and yet they are so invested in living membrane, that additions of carbonate of lime are constantly made to each. By this arrangement, the animal within grows without inconvenience to itself, which it could not do, if the integument or dwelling was composed of one piece.

The forms of the Echinoderms differ much among themselves, and yet it is apparent that they all belong to one type, and are constructed upon one plan. One of the most aberrant of this type is the sea cucumber, (Holothuria,) which is a firm fleshy bag, destitute of plates, composed of carbonate of lime. In another upon our coast, we find the *star-fishes* with five arms extending from a common center; and in another, the globular *sea-urchin*, in which the five arms are folded and soldered together so as to form a ball. Another interesting form has the stellate type, but differs considerably from the star-fish, and most strikingly in the fact that the stellate head is supported on a jointed foot-stalk. These are called *Encrinites.*

These different families have a special geological interest. The last for example, the Encrinite, lived in the earliest periods of the planet, and are known principally in the oldest palaeozoic rocks. In the lower silurian system, beds are often composed mainly of their disarticulated remains. In modern rocks and seas, they are unknown. On the contrary, the star-fishes without pedicels or jointed supports, are known mostly in modern rocks, only two or three species being known in the earlier formations. Now, the sea-urchins, or the globular forms of this class, lived in great numbers in the Mesozoic or Jurassic period. This type or form has come to us, though none of the species of the Mesozoic period live in our present seas.

I have spoken of the complicated structure of the star-fishes and the provision which has been made for their growth, both of which are worthy of our highest admiration. But nature had not exhausted all her resources when she had provided for their growth and made them the most beautiful objects in the seas. She has in this elaborate structure made their ornamental work subordinate to their instruments of locomotion and reproduction. The flowers which are sculptured upon their integuments form a part of their organs for moving from place to place. These flowers which represent the five petals of a rose, are formed by punctures through the outer envelope. Through them the urchin protrudes fleshy suckers or tubes. If, for example, a sea-urchin is placed in a glass filled with sea-water, it is soon seen to protrude a multitude of slender fleshy threads, each of which is tipped with a little knob. These soon come in contact with the glass to which the knob adheres, on the principle of an exhausted receiver. By means of this adhering apparatus, it moves itself forward or backward. In technical language, the surface from which these fleshy threads protrude, are called *ambulacral areas*, and the spaces between, *interambulacral areas.* Nothing can be seen of these threads when the animal is dead. All its soft parts are strictly encased in a box of hard shell substance, which has received the name of Test, or Shell. The patterns of these different areas vary in form and proportion, and hence are used as characteristics of genera and species. The test is also covered with spines of different forms and sizes. These, too, are formed after different patterns, their shafts being sculptured differently in every species. Their spines, and the mode they are attached to the shell, the character of their surfaces, the position of their oral and excretory orifices, furnish the characters upon which the families, and lesser subdivisions of this class are founded.

FAMILY CIDARIDAE.—CIDARIS MITCHELLII.—N. S. (Fig. 237.)

Test thick, circular or turban shaped; flattened above and below; ambulacral areas narrow, and provided only with minute tubercles, in double rows, and three in each; interambulacral areas nearly four times as wide as the former, and furnished with two distinct rows of large primary tubercles, with about eight in a row, including the smaller ones upon the disks; tubercles perforated; inner rim surrounding the tubercle, smooth; outer, bearing small subordinate spines, giving it a crenulated appearance; miliary zones wide, and covered with small close set unequal granules; poriferous zones, unigeminal, and separated by nearly plane ridges; spines unknown; apical disk unknown; mouth opening, appears to be large, but too much broken to determine its characters.

FIG. 237.

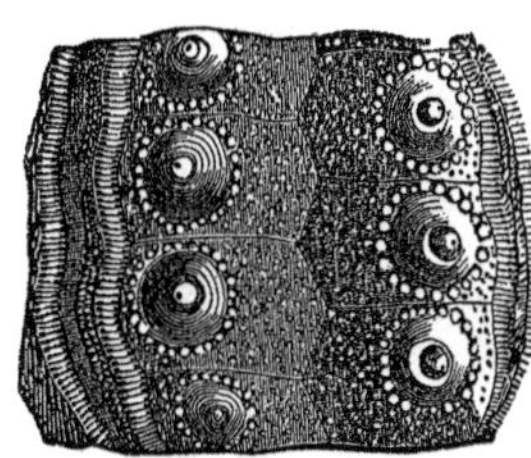

Belongs to the eocene, and accompanies the remains of the Zeuglodon.

Dedicated to the lamented Prof. Mitchell of the University of Chapel Hill.

CIDARIS CAROLINENSIS.—N. S. (Fig. 238.)

Test rather thick, circular and somewhat oval. Ambulacral areas narrow; somewhat undulating, supporting two rows of small tubercles with two in a row, and interspersed with minute ones, which appear in some places to be arrayed in subordinate rows; interambulacral areas wide, covered with small subequal and rather prominent tubercles, among which minute granules are scattered; area about four times as wide as the former; plates pentagonal, supporting two rows of large perforated primary tubercles, surrounded by plain circular zones; miliary zone concave or depressed. Poriferous zones narrow; pores unigeminal; outer oblong; the inner circular; margin of the small plates between them marked with an elongated depression. The upper and lower sides crushed.

Fig. 238.

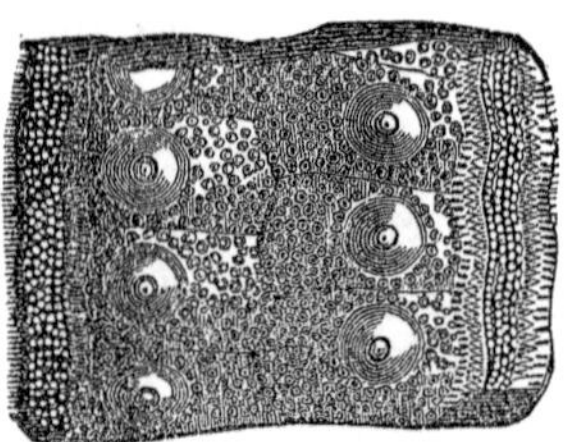

Belongs to the eocene, and accompanies the former.

Figure 105 represents the jaws of an Echinoderm, p. 246. The separate pieces of the test and jaws are quite common in an eocene bed in Craven county. They belong to the upper part of the bed, and seem to be confined to a space about two feet thick.

FAMILY CIDARITAS.—ECHINUS RUFFINII.—ED. FORBES. (Fig. 239.)

"Body sub-depressed; ambulacral and interlambulacral; plates with several primary tubercles on each closely ranged, having circles of secondary tubercles surrounding their bases; rows of pores very oblique, with three pair of pores in each row, the uppermost distant from the other two. Beneath concave; mouth broad; widely notched opposite each avenue." Ed. Forbes.*

Fig. 239.

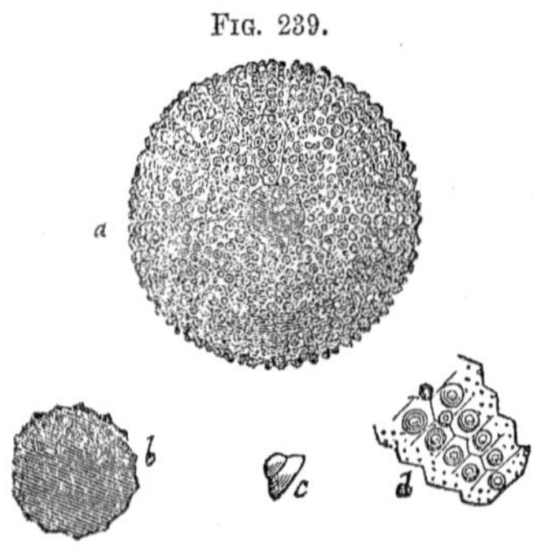

* Journal Geological Society, vol. I, p. 426.

Found in the miocene beds. Four views, *a*, Echinus Ruffinii, viewed from above; *b*, mouth; *c*, spinegerous tubercles; *d*, ambulacral plates, and arrangement of pores: *a*, *b*, natural size, *c*, *d*, enlarged.

FAMILY CLYPEASTARIDAE.—ECHINOLAMPAS APPENDICULATUS.—N. S. (Fig. 240-'1.)

Fig. 240-'1.

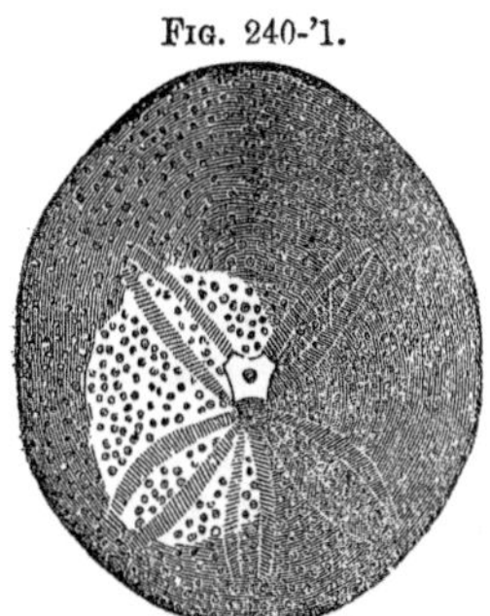

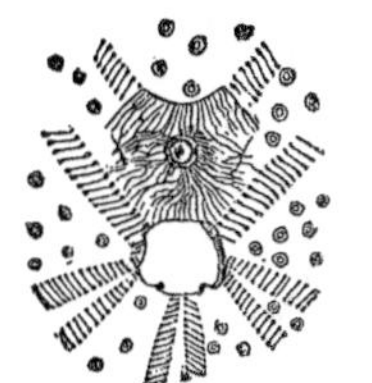

Test thin; body oval, depressed; margin thick or rounded; somewhat elongated, wider anteriorly than posteriorly; ambulacra narrow, open at their extremities; sub-petaloid; pores connected by furrows; mouth transverse; excretory orifice horizontal, marginal; madriporiform plate excentric; apical disk occupied by a sub-cordate sculptured plate, furnished with a pentangula opening, in the centre of which there is a pore; areolæ more numerous below than above; area around the mouth inflected.

ECHINOCYAMUS PARVUS.—N. S. (Fig. 244.)

Test small, oval, with rounded sides; avenues dorsal; mouth sub-central, rounded, large, with a crenulated margin; vent between the mouth and hinder margin; genital pores apparently four. Figure natural size. The mouth is large in proportion to the size of the body and the vent is situated half way between the mouth and margin. Eocene of Craven.

Fig. 244.

SCUTELLA LYELLII.—(Fig. 246.)

FIG. 246.

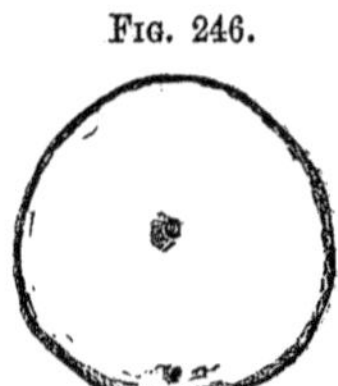

Shield small, sub-circular, flat, scarcely convex above; below slightly concave; ambulacra open towards the margin and terminating in four pores; in that direction mouth small; vent near the margin. Eocene, Wilmington.

SCUTELLA.—(Fig. 247-'8.

Figures 247-'8 represent a common fossil of the eocene of Craven county. 247 inferior face, showing the relative

FIG. 247-'8.

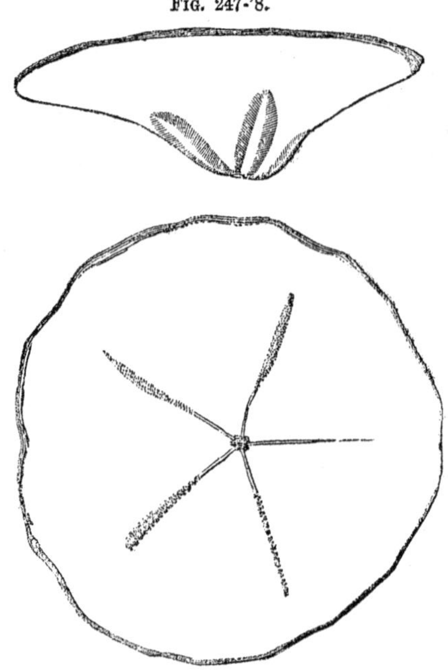

position of the mouth and excretory orifice. Figure 248 is profile view of the same. The apical summit is before the genital. Since its discovery no opportunity has been furnished by which I could obtain a comparison with the forms already known and described by the palaeonlologists of this country. Wadsworth's eocene marl, Craven county.

FAMILY SPATANGIDAE—GONIOCLYPEUS SUBANGULATUS.—N. G.
(Fig. 242.)

Test thick, sub-conical, covered with small spines, anterior and posterior areas somewhat unequal; margin and base

FIG. 242.

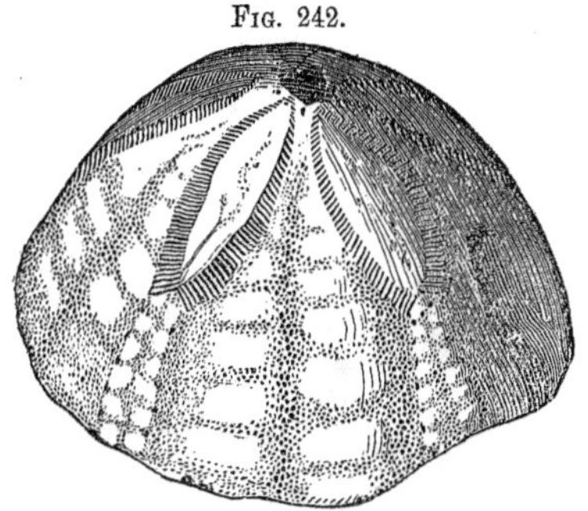

FIG. 243.

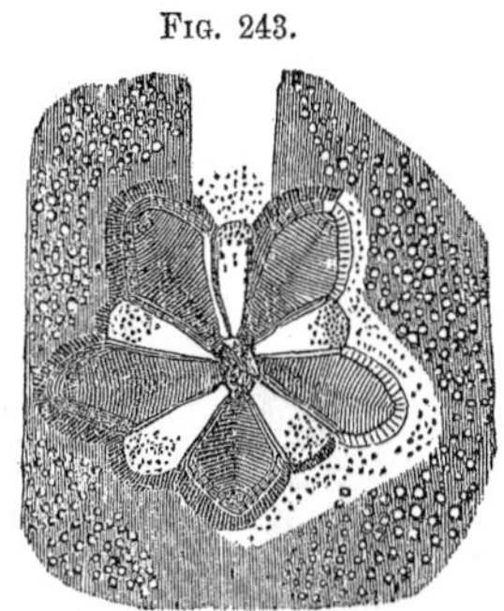

somewhat pentangular; posterior or anal orifice lateral, or upon the superior face; interambulacral area grooved, with the continued area beneath projecting; interambulacral areas sub-angulated; mouth rather narrow or small, central; peristome angular, and surrounded by five angular prominences, which terminate in the interambulacral areas, between which is a rosette, perforated by seven pairs of pores, with three odd ones at the end of each petal; ambulacra petaloid and closed; the prolonged zone provided with alternating pores as far as the base; pores connected by oblique grooves; interambulacral wide; plates large, and nine or ten in a column.

Figure 243, rosette enlarged.

OBSERVATIONS.—The ambulacral areas are narrow, but the poriferous zones are rather wide; and the interambulacral areas are about four times as wide as the ambulacral. The genital plates are indeterminate, but the pores are large and the occular small, and appear to be mere indentations; buccal area ornamented with a rosette; petals transversely wrinkled; pores elongated; the anterior lateral plates appear to have eleven pairs of pores instead of seven. The genus is closely related to Cassidulus of Lamark, but the pores are united by grooves. Eocene, Wardsworth marl, Craven co.

AMPHIDETUS VIRGINIANUS.*—FORBES. (Fig. 245.)

"Body broadly ovate, elevated and truncate posteriorly; back oblique; dorsal impression lanceolate; scutab area very slightly excavated; ambulacral spaces broad, triangular, depressed; interambulacral spaces slightly convex; anteal furrow broad and shallow, sides slightly gibbous; sub-anal impressions broadly ob-cordate; post-oral spinous space broadly lanceolate.—Edw. Forbes."

FIG. 245.

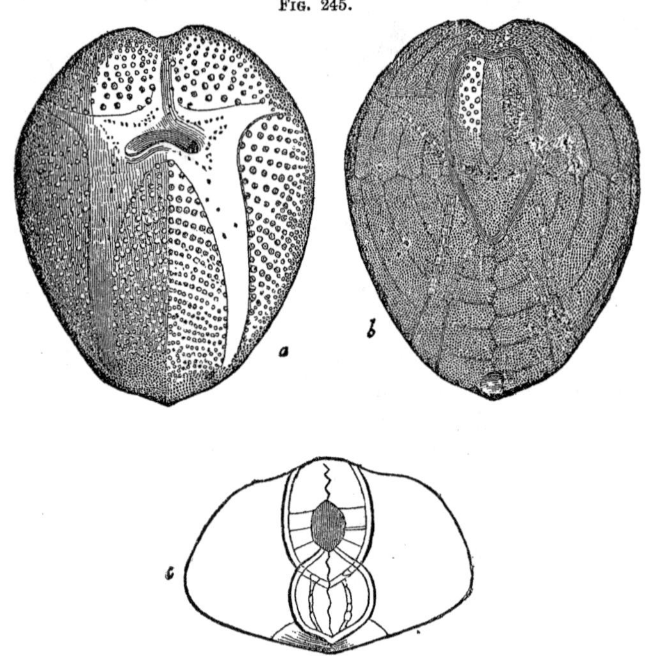

a, lower area; *b*, upper area; *c*, posterior area, showing the relation of the sub-anal impression. Usually found in fragments in the miocene of North-Carolina.

* Journal of the Geological Society, Vol. 1, p. 425.

ORDER CRINOIDEA.—MICROCRINUS CONOIDEUS.—N. G.

(Figs. 246 & 247.)

FIG. 246. FIG. 247.

Body conical; sub-pentangular at base; areas five, oblique; pores six or seven to each, alternating and arranged in rows, separated by a ridge; apical pores five, base wide; beneath concave; concavity intersected by five bars, which descend and meet in the center; spaces between, triangular, terminating above in the apical pores.

Figure 247 shows the base with the intersecting bars and triangular spaces between.

I am unable to determine whether the head is supported on a foot-stalk; the joints of a crinoid, however, are numerous in the marl in which this curious species is found.

Eocene of Craven county, and associated with Echinocyamus Parvus.

BRYOZOA.—LUNULITES DENTICULATA.—(Figs. 248 & 249.)

"Conical; cells inalternate, oblong externally, interior conical, nearly vertical to the two surfaces of the polypidom; margin of the cell in its immature state open and denticulated; when mature, covered; mouth near the distal extremity; semicircular when imperfect, circular when perfect; gemmuliferous chamber at the distal end of the cell, opening round, concave surface furrowed, irregular and minutely granulated." * Miocene, and common to most of the beds upon the Neuse and Cape Fear.

FIG. 248. FIG. 249.

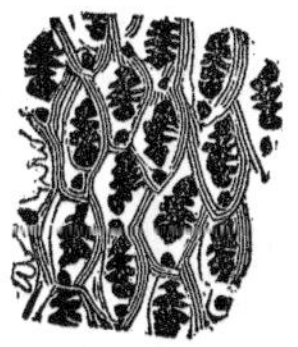

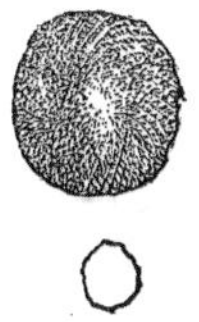

Fig. 249, enlarged view of the fossil, showing the arrangement of the cells, and the small Figure its natural size.

LUNULITES CONTIGUA.—Figs. 250 & 251.

The figures exhibit casts of the concave surface of the

* Lonsdale, miocene corals from N. America, Journal Geol. Society, vol. 1, p. 503.

coral. Fig. 251, cast of the concave surface natural size; Fig. 250, magnified view of a portion of the surface. Eocene Wilmington.

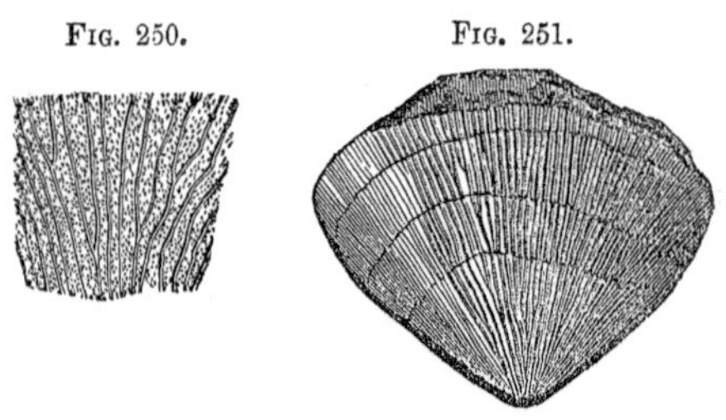

Fig. 250. Fig. 251.

LUNULITES OBLONGUS.—N. S. (Figs. 252 & 253.)

Polypidom small, conical; cells arranged along a straight line, from the base to the margin; open cells show that they are nearly quadrangular; the closed cells do not show an orifice; there is a simple film spread over the cell, and the margins are simple and unlike denticulata. Fig. 253, greatly enlarged view of the cells; small figure shows the natural size of the fossil.

Fig. 252. Fig. 253.

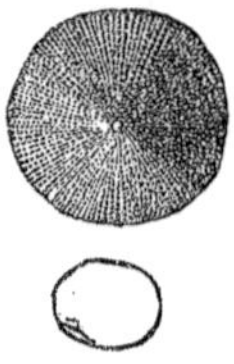

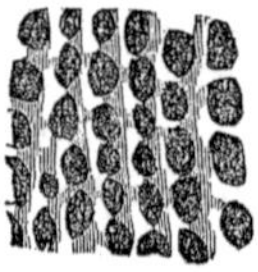

DISCOPORELLA UMBELLATA.—(Figs. 254 & 255.)

It is impossible to discover any difference between our Discoporella and that of the miocene of France; the cells have two orifices at opposite acute angles, and the same arrangement of cells. Fig. 255 greatly enlarged. This figure, however, fails to give a clear and correct view of the fossil. A reference therefore, to Pietet's Pl. XC, page 15, is necessary.

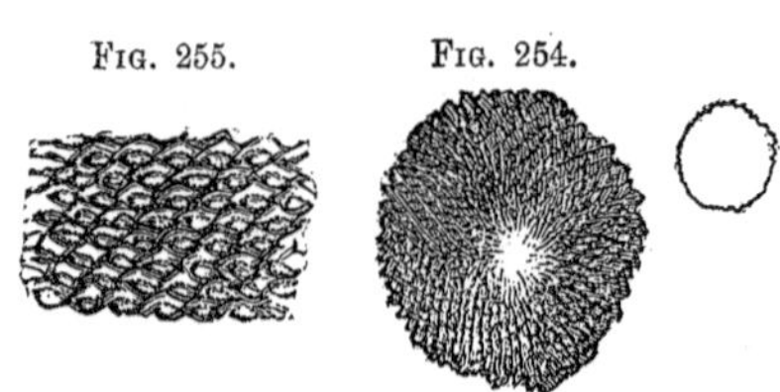

Fig. 255. Fig. 254.

The small lunulites begin to form at the apex, and for this

purpose they attach themselves to a grain of sand, which will generally be still found at the point of growth; some of the miocene ones are nearly half an inch in diameter.

POLYPAIRA.—ASTRAEA BELLA.—(Fig. 256.)

The stars are polygonal, variable, rather deep, lamellar lamellae twelve, with alternating ones, denticulated, contiguous, or separated by their partitions.

FIG. 256.

Common in the miocene incrusting shells, and various bodies found in a marl bed.

ASTRAEA.—(Fig. 256a.)

Irregularly branched; stars deep and rather distant, though in some places contiguous as in the Bella; intermediate spaces without pores, but bordered by lines to which the lamellae extends; lamellae denticulated, as in A. Bella, and provided also with the same number, and similarly arranged. Miocene.

FIG. 256a.

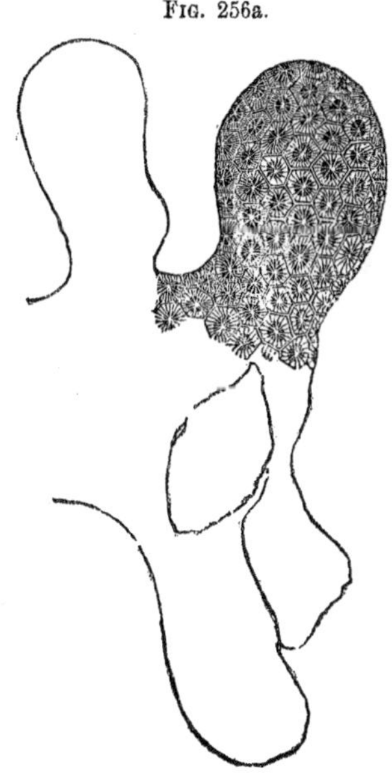

The foregoing sketch of the fossils of the marl beds of the eastern counties, is far from being complete. Numerous species still remain unnoticed and undescribed. It seemed to be desirable, however, on many accounts, to illustrate some of the interesting contents of these beds, which are truly the only historical mementoes which now remain to us of the ages during which they lived. It will appear, on examination, that I have placed by far the largest number of species in the miocene. I have thus placed them because the shell marl beds contain so large a number of the acknowledged miocene fossils of Virginia; and besides, there are many which replace miocene fossils of Europe.

In conclusion, it is due to myself to remark, that the circumstance under which many of the determinations have been made, rendered it impossible to consult authorities, and hence it may turn out that many species which have been marked as new, will prove to be old ones already described. The course I have pursued may have been injudicious, and hence may open the way for censure; still, under the circumstances, I deemed it the best I could pursue.

ADDITIONS AND CORRECTIONS.

I.—Figures and Names of Species:—

Page 205, for *otololite* read *otolite.*

" 242, fig. 90, read Galeocerdo Egartoni.

" 241, fig. 84a is Sphyrna denticulata.

" " " 82a and 83a, Galeocerdo contortus.

" 243. It is possible *Trygon*, fig. 94, should be referred to Myliobatis.

" 245.—Fig. 105 is the valve of the genus *Scalpellum* of the class *Cirripedes.*

" 261, fig, 139.—This is not Erato laevis, but is closely allied to E. Maugeriæ, of the coralline crag.

" 268.—Fig. 159 resembles *Cerithium adversum* of the English crag.

" 290.—For *Lucenidae* read *Lucinidae.*

" 291.—Place a period before Brugiere.

" " —For *Pennsylvania* read *Pennsylvanica.*

" " seucond line from bottom, for *multihineata* read *multilineata.*

" 292.—For *Venerida* read *Veneridae.*

" " —For *Tridaenoides* read *Tridacnoides.*

" 293.—For Cribrari read Cribraria.

" " second line from bottom, for *pramagna* read *permagna*; for *meltastriata* read *metastriata.*

" 294.—For Cytherca read Cytherea.

" " For reporta read reposta.

" 295.—For Artemes read Artemis.

" 296.—Fig. 224 shows the hinge of Artemis tranversus; and read Artemis for Artemus.

" " sixth line from bottom, for Tellimidae read Tellinidae; and ninth line, for Tiphonal read Siphonal.

" 297.—For P. Sammobia read Psammobia.

" 306.—For Cidaritas read Cidarites.

" 307, second line from top, for Spinigerom read Spinigerous.

" 311.—Bryozoa should have been placed under an independent head, as a subdivision of Molusca and not under Radiata.

Certain figs. have been placed wrong side up, particularly Scutella. fig. 247—'8.

In the Eocene of Craven county, I have found the palatine teeth of the Saurodon, or Saurocapalhus, and also fragments of a Xiphioid fish, as the prolonged premaxilary of a sword fish.

Retinasphalt occurs in the marl of Duplin county.

www.ingramcontent.com/pod-product-compliance
Lightning Source LLC
LaVergne TN
LVHW020219110826
845151LV00003B/760